KEVIN DUTTON

BLACK-*and*-WHITE THINKING

Kevin Dutton is a research psychologist and a fellow of the British Psychological Society who has spent the best part of the last twenty years at the Universities of Oxford and Cambridge. He is the author of the acclaimed bestsellers *Flipnosis: The Art of Split-Second Persuasion* and *The Wisdom of Psychopaths: What Saints, Spies, and Serial Killers Can Teach Us About Success*, the latter of which was featured in *The Best American Science and Nature Writing*. His work has been translated worldwide into more than twenty languages, and his writing and research have been featured in *Scientific American, The Guardian, Psychology Today, The New York Times, The Wall Street Journal*, and *The Washington Post*, among other publications. Alongside his academic commitments, Dutton also regularly consults in elite sports, business, and military sectors. Follow him on Twitter at @TheRealDrKev.

BLACK-*and*-WHITE

THINKING

BLACK-*and*-WHITE THINKING

The Burden of a Binary Brain
in a Complex World

∶∶

KEVIN DUTTON

PICADOR

FARRAR, STRAUS AND GIROUX

New York

Picador
120 Broadway, New York 10271

Grateful acknowledgment is made for permission to reprint the following material:
Extract of a letter to the *Daily Mail* published on November 11, 1914,
by permission of the *Daily Mail*.

Owing to limitations of space, illustration credits may be found on page 387.

The Library of Congress has cataloged the Farrar, Straus and Giroux
hardcover edition as follows:
Names: Dutton, Kevin, 1967– author.
Title: Black-and-white thinking : the burden of a binary brain in a complex world /
   Kevin Dutton.
Description: First American edition. | New York : Farrar, Straus and Giroux, 2021. |
   "Originally published in 2020 by Bantam Press, Great Britain." | Includes
   bibliographical references and index. | Summary: "How the evolutionary history of
   the human brain explains our tendency to sort the world into black-and-white
   categories"—Provided by publisher.
Identifiers: LCCN 2020039385 | ISBN 9780374110345 (hardcover)
Subjects: LCSH: Categorization (Psychology)—Popular works. | Social psychology—
   popular works. | Human behavior—Popular works.
Classification: LCC BF445 .D67 2021 | DDC 150.19/5—dc23
LC record available at https://lccn.loc.gov/2020039385

Paperback ISBN: 978-1-250-82945-0

Our books may be purchased in bulk for promotional, educational, or business use.
Please contact your local bookseller or the Macmillan Corporate and Premium
Sales Department at 1-800-221-7945, extension 5442, or by email at
MacmillanSpecialMarkets@macmillan.com.

Picador® is a U.S. registered trademark and is used by Macmillan Publishing Group,
LLC, under license from Pan Books Limited.

For book club information, please visit facebook.com/picadorbookclub or
email marketing@picadorusa.com.

picadorusa.com • instagram.com/picador
twitter.com/picadorusa • facebook.com/picadorusa

In the beginning, when God created the universe, the earth was formless and desolate. The raging ocean that covered everything was engulfed in total darkness, and the Spirit of God was moving over the water. Then God commanded, 'Let there be light' – and light appeared. God was pleased with what he saw. Then he separated the light from the darkness, and he named the light 'Day' and the darkness 'Night.' Evening passed and morning came – that was the first day.

GENESIS I: I-5

Contents

BLACK-*and*-WHITE

THINKING

Introduction

∷

He's like a man with a fork in a world of soup.

NOEL GALLAGHER

O N ONE PIECE OF PAPER are scribbled the words 'real life'. On the other, the word 'fantasy'. Each piece is sellotaped to a jam jar next to the cash register and in the middle sits a picture of Freddie Mercury. The jars are rammed three-quarters full with coins and bills. Which, it turns out, don't take long to accumulate. By the time I finish my entrée both jars have been emptied and two new labels doled out. 'Kittens' on one side. 'Puppies' on the other. Not quite in the same league as the lyrics to 'Bohemian Rhapsody' perhaps. But effective, nevertheless. The resonant chink of metal on glass continues.

I'm curious.

I'm sitting in a café in San Francisco, where I've just spent the last fortnight talking to three of the world's leading experts on black-and-white thinking – the shallows of the binary brain. With time to kill before heading home to Oxford I've ventured into Haight-Ashbury to reflect on things. I order some tacos and decide to ask the waitress what the deal is. She smiles.

'We keep changing the labels,' she tells me. 'Five, six times a day. Before, when we just had the one jar and there was no choice, the tips were very slow. But if you give customers the option – kittens or puppies – they're way more generous. I don't know why. It's just a bit of fun, I guess.'

I'm not so sure.

Before I leave, I hang around the cash register waiting to pounce. Two women in their early twenties hesitate, giggle, and then break rank. One goes for 'kittens', the other 'puppies'.

Why? I ask.

'Cats don't need you,' says one woman. 'Puppies do.'

Her friend shakes her head. 'That's exactly why I prefer cats! You never need to take a cat out for a walk. But you can't very well not walk a dog. Where's the fun in that when it's cold, dark and raining—'

Puppy Woman cuts her off. She's not letting this one go. 'Which is why dog people are more friendly,' she protests. 'When you take your dog out for a walk you meet other dog walkers and you start to get to know each other.'

The to and fro continues for a while until they go off, bickering. The waitress comes over and rings up another sale.

'See,' she says. 'I told you. People enjoy it when they have to make a choice. And, in the process, they go away happy.'

I nod. But I can't help wondering about some of those other choices she mentioned. What other options were customers presented with when they came to settle their bills?

She shrugs. 'Apple or Microsoft,' she says. 'Spring or fall. Bath or shower . . .'

She tails off as she zips away to another table. But the list, I'm guessing, is endless. Because, actually, there are any number of binary ways you can divide people. Any number of opposing axes of preference you can fashion from our composite identities.

My mind goes back to a piece I'd read recently in some newspaper or other. Apparently, Facebook currently had some seventy-odd different categories for gender identity alone. And there were getting on for something like 4,000 different music genres on Spotify. In a murky world of fuzzy, blurry boundaries and ever-increasing border-creep we jump at the chance to categorize ourselves unconditionally. To nail our colours definitively to the mast. Especially when those colours happen to be clean, simple, and psychologically unchallenging.

As the jam jars attest, we'll even pay for the privilege.

We live in a divided world. Everywhere we look, there are lines. Countries, most noticeably, have borders. On one side is 'us'. On the other, 'them'. Cities have districts and neighbourhoods. But in everyday life, the lines that we draw are endless. We draw lines based on gender. We draw lines based on race. Here in the UK, we even draw lines based on Europe. In or out.

Our brains come equipped with a formatting palette. We're hard-wired to draw lines by our rich evolutionary past. But how can we be sure that the lines we are drawing are accurate? And how do we know where to place them? The answer, quite simply, is: we can't. We have no way of knowing, no means by which to be certain, that the lines we are drawing are true. And yet still we are compelled to draw them. Because the world is a complicated place and lines make stuff easy and doable. And 'doable' is something we crave.

By way of example let's take student grade point averages (GPAs). In academia, degree classes are awarded on the basis of where a candidate's final year average falls within the spectrum of a preset range of marks. On the one hand, this is perfectly reasonable. But on the other, statistically impressionist. Does it really make sense to talk of one student having a 'first-class mind' and the other not, if one averages seventy and the other sixty-nine?*

This is a question that doesn't just ooze philosophical appeal

---

* In the UK, a score of 70 per cent or higher is equivalent to an 'A'.

but which has practical implications, too. Opportunities exist for students with first and upper-second class degrees that do not exist for those with lower classes. A point here or there at or near a grade boundary can spell the difference between continuing on to a higher degree or, academically speaking, journey's end. Dreams can be dashed. Career prospects ruined. Doors open and close depending on where we draw our lines. Sometimes, quite literally. As part of the strategy for dealing with the spread of Coronavirus in the March of 2020 the UK government urged all people over seventy to remain within the confines of their own homes to protect themselves against the disease. Over seventy when it comes to exams, doors open. Over seventy when it comes to Coronavirus, doors close.

But we have, as we say, to 'draw the line somewhere'. And we do. Storms, drugs, prisons, terror threats, pandemics* . . . you name it, we categorize it. We use numbers, letters, colours, anything that comes to hand. Because a line drawn is a decision made. And life is full of decisions.

Black-and-White Thinking, then, is about order. Or rather, it is about the illusion of order. It is about how the lines we draw in the desert of unbroken reality evaporate like cognitive mirages the closer we zone in on their chimeric, ephemeral forms. Back in the days of our prehistoric ancestors our brains were fresh out of the box. They ran smoothly, efficiently, and were exquisitely fit for purpose. Were our ancient forebears to have stumbled upon a snake beneath a rock,

---

* At the time of writing the Coronavirus (COVID-19) is dominating the news headlines. In response, Public Health England have divided those countries with implications for returning travellers or visitors arriving in the UK into two categories based on the risk of infection. *Category 1:* 'Travellers should self-isolate, even if asymptomatic, and call NHS 111 to inform of recent travel. Go home or to your destination and then self-isolate.' *Category 2:* 'Travellers do not need to undertake any special measures, but if they develop symptoms they should self-isolate and call NHS 111.' (From www.gov.uk)

uncertainty over whether or not it was dangerous simply didn't apply. You got out of there. Fast. The same went for tigers in bushes. And crocodiles in reeds. OK, it *could* be the wind. But better to reflect on that fact from a safe and comfortable distance, well out of range of the sharp, the pointy or the poisonous.

In other words, the overwhelming majority of the decisions that our primeval forefathers would have made during the course of their everyday lives were likely to have been binary. Black and white. Either/or. And with good reason. The decisions being made were often a matter of life and death. Flash floods. Tornadoes. Lightning strikes. Landslides. Avalanches. Falling trees. These things come out of nowhere. They happen in the blink of an eye. Chin-strokers and navel-gazers usually weren't around for too long.

Today, however, the survival game has changed. The mental shortcuts that ensured our ancient ancestors remained ahead of the evolutionary curve can, like those very snakes and tigers they evolved to help us evade, themselves come back to bite us. The evidence is all around. Just ask the South African female middle-distance athlete Caster Semenya, who has come under fire for naturally occurring elevations in her testosterone levels. Or Caitlyn Jenner. The relentless pressures of ever-increasing nuance, of social, psychological and informational complexity, demand larger scale maps of an endless, seamless reality with finer lines and fewer, less prominent creases. The clunky delineations of yesteryear quite literally no longer cut it.

To see how the demands on the brain have evolved, let's go back to where it all started and consider the single-celled amoeba. The aim of all living things is survival and reproduction. In the shallow, aquatic photosphere that represents the sum total of this rudimentary organism's entire universe, fluctuations in heat, the availability of food, and ambient luminosity – light and dark, black and white – constitute, in effect, the only three categories essential for its survival. Such changes in the amoeba's external environment are mirrored by internal changes within the organism's cell membrane enabling it to move either towards proximal sources of food (i.e. glucose) or away from noxious stimuli. As one amoeba I spoke to put it: 'No need for a dimmer switch in this pond, pal. Romantic, candlelit soirees aren't really our thing. If it suddenly goes dark it means there's something, quite literally, in the air and we make ourselves scarce. We prefer to keep it simple.'

A noble, if Spartan philosophy.

But simplicity isn't for everyone and under the tutelage of natural selection came enterprise, innovation and emerging technological know-how. The appearance of multicellular organisms heralded the arrival of the world's first nervous systems in the form of primitive nerve nets or ganglia – tangled up clusters of nerve cells. Just as with their single-celled prototypes, their purpose and function was simple: to facilitate the detection of extracellular stimuli, only this time, a billion or so years later, through quicker, more efficient means.

Scroll forward to the present and our painstaking psychophysical makeover is complete. We humans now come equipped with at least six primary senses and probably a whole lot more: five of them designed to monitor goings on in the outside world – sight, touch, smell, taste and hearing – and one, proprioception, geared towards maintaining a stable internal equilibrium through a constant stream of coordinated neural press releases on movement and body position.

Of these six basic senses vision is by far the most dominant, accounting for around 70 per cent of all of our sensory input. And with good reason. During the course of our evolutionary history, an accurate representation of the size, colour, shape and motion of other living things around us would've been essential to our survival – as, just as it still is for the amoeba, would sensitivity to brightness and contrast. To our prehistoric forefathers the sudden appearance of a shadow on the wall of a cave, on the surface of a body of water, or on a parched patch of open, sun-bleached savannah precipitated two elementary considerations. One: is this something I can have for dinner? Two: is it something big, strong and sharp that wants to have *me* for dinner?

Light and dark. Black and white. Three and a half billion years from amoeba to Neanderthal. Not, upon reflection, an inordinate amount of change.

Then, of course, everything *did* change. Overnight, pretty much, on a paleontological timescale. With the emergence of consciousness and the advent of language and culture, the goalposts of human advancement didn't just move. They were shipped off and reassembled on a different evolutionary playing field entirely. Suddenly, black and white – light and dark – were yesterday's news. And grey was the colour to think in. Dimmer switches became indispensable; an essential piece of neurophysiological kit to help us reason in hues and shades rather than anticipate in binary monochrome.

Except there was a problem. A big one. There weren't any. Dimmer switches didn't exist. They *might* have been the next big thing. *Could* have been the new fight or flight. But the market had changed at such a lightning speed that the eggheads at Natural Selection hadn't got round to manufacturing them. They hadn't gone into production. Still haven't, in fact.

Which leaves us in a bit of a predicament. Up grey creek with just the simplest of black-and-white paddles. We group, we label, we pigeonhole our way to making irrational and suboptimal decisions all because our brains grew too big too fast. Because they got too clever, too quickly, too soon. We categorize rather than graduate. We polarize rather than integrate. And we exaggerate and caricature variation as opposed to emphasizing and accentuating similarity.

Take wine, for example. Many of us, I imagine, would be able to distinguish red from white in a blind taste test. And some, perhaps, a cabernet sauvignon from a pinot noir. But how would you fancy a shootout between two rarefied, sophisticated and highly complex Bordeaux: say, a *Château Lafite Rothschild En Primeur 1982*

and a *Château Mouton Rothschild En Primeur 1995*? Hmmh, maybe
not. More worrying, of course, is our blindness to differences
between *people*. Categorizing a *Château Lafite Rothschild En Primeur
1982* as 'red' or a *Jacques Prieur Montrachet Grand Cru* as 'white' is
hardly going to trigger the Apocalypse. However, categorizing all
Muslims as Islamic State (Isis)-supporting radicals just might.

The deadly Darwinian irony is lost on us at our peril. Far from
being essential for survival, binary cognition could one day seal our
fate. Sooner rather than later if the polarizing strategies of terrorist
organizations like the aforementioned Islamic State or populist
brands of political fundamentalism such as that oft-peddled by
the Trump administration continue to steal a march on sense and
reason.

Yet aren't all of us equally to blame? Everywhere we look, our
chequerboard minds divide us. No more so than in the ephemeral,
click-and-run world of social media. If someone puts forward a
claim or belief that we don't wholeheartedly agree with, then what
is our first reaction? Aren't we naturally inclined to argue the polar
opposite? (See Figure 0.1 opposite.)

In the pages that follow then, we begin by looking at the importance of categories and categorization – the infinite cognitive Lego
set with which we assemble and construct reality – in everyday life
and reflect upon the fact that without it we wouldn't be capable of
making even the simplest of decisions. As the book takes shape,
we'll discover how we begin to draw lines from a very early age;
how our ability to categorize is an instinct like language and walking, an evolutionary adaptation buried deep within our brains
rather than something we acquire completely and utterly from
scratch. And we'll discover the reason behind this: how the cognitive imposition of such 'false clarity' satisfies a primeval human
need for order. For the simplicity and 'doability' we crave so deeply.
For distinctions, dichotomies and borders. But how, when the lines
we draw are between people rather than pictures, distinction and

BLACK	WHITE	GREY
SPORT: Only Olympic gold medals should be celebrated. Silver and bronze ultimately denote failure.	All Olympic medals should be celebrated. Any podium finish is an achievement.	On some occasions lesser medals are a cause for celebration. On others not. A nailed-on favourite who slips down to silver due to a lack of concentration might be a case of the latter. A rank outsider who performs way beyond expectation might be a case of the former.
POST-BREXIT IDENTITY CRISES: I am British.	I am European.	You can be both. You don't need to give up feeling British in order to feel European. And vice-versa. Identity is contextual. When England play another European country at football an England supporter will feel 'English'. But should that same supporter find himself in a remote outpost in the far-flung reaches of another continent he may well feel predominantly 'European'.
POLITICS & NATIONAL SECURITY: Terrorists are mad.	Terrorists are sane.	Many terrorists may *behave* 'madly' but are actually of sound mind. Two people may perform the same action but for vastly different reasons. The processes of radicalization can turn a perfectly normal individual into a psychological avatar of a lunatic leading them to perform insane acts from a perfectly rational perspective.

Figure 0.1: Three examples of black-and-white arguments that have recently appeared in the British media (and their grey alternatives).

dichotomy can very quickly metastasize into division, discrimination and discord.

Take, for instance, in the cultural sense, black and white itself.

Imagine that we were to arrange all of humanity in a line according to skin colour, from black to white, and were to make our way along it. At no definitive point as we negotiate its length would we be able to discern where a black person ends and a brown one begins. Or where a brown person ends and a tan one begins. Or where tan ends and white picks up the thread. In terms of skin colour, those standing shoulder to shoulder with each other would actually be all but identical. We'd be faced with a continuum of indeterminate intermediates. Black and white, in a racial or an ethnic sense, would simply cease to exist.*

Yet race is one of the hottest topics around.

As we continue our look at the brain's innate, all-encompassing categorization instinct we stumble upon other hazards too. Freddie Mercury was apparently fond of asserting – when he wasn't otherwise engaged pondering the differences between real life and fantasy – that if a thing was worth doing it was worth overdoing. Far be it from me to argue with Freddie over the dos and don'ts of putting on a show, but when it comes to how we categorize he couldn't have been any more wrong. Too many options and our brains don't know which way to turn. Too few and we veer into militancy.

As an example of the former problem, take film categories. Netflix currently lists over 76,000 meticulously curated sub-genres of movie ranging from 'Psycho-Biddy' (films about feuding grannies engaged in bitter rivalries with each other in their haunted mansions) to 'Sea Creatures Playing Sports' (films depicting mutant sea creatures engaged in various sporting activities, including a giant crab playing a soccer goalkeeper in *Kani Goalkeeper* and a giant Calamari playing a wrestler, in, no prizes for guessing, *The Calamari Wrestler*).

As an example of the latter, take Isis. The 'grey zone' is a term specifically conceived of by the organization to describe a world in

---

* This is not to imply that race and ethnicity are *just* about colour but rather that colour plays a key role not just in racial *identity* but in racial *identification* too.

which Muslim and non-Muslim might live together. It is an abomination, a blasphemous blurring of the brutal, binary line that delineates the frontier between two monolithic meta-categories. Them and the infidel. The just and the unjust. Ones and zeros. Bums and heroes. No wonder that their flag bears just the two colours, black and white.* That is no coincidence. And no wonder that a city such as London, indeed any city, is anathema to them. Cities are all about confluence; a coming together of hopes and dreams, passions and bugbears, colours and creeds; a muddying of the existential waters. The multi-faith, pan-ideological heartbeat of a modern thriving metropolis, be it London, Paris, New York or Beirut, rocks Isis – and other organizations like it – back on their heels because it suggests that we humans might just be able to make a go of it together. Get along. Those who are Muslim. And those who are not. Shopping in the same markets. Watching the same movies. Walking the dogs and playing with the kids in the same parks and playgrounds.

But fanaticism stands for none of it. Zealotry demands the rapid and complete rejection of even the slightest hint of ambiguity. To Isis, trans-ideology of any denomination represents decadence, depravity and godlessness of infinite, apocalyptic proportions. It has to be one or the other. Either/or.

The solution, of course, lies somewhere in the middle, and we'll discover that between goalkeeping crabs and calamari wrestlers on

---

* The Black Banner or Black Standard is a flag depicting a uniform black background flown by Muhammad in Muslim tradition. It also constitutes an important symbol in Islamic eschatology heralding the advent of the Mahdi (or 'guided one') who, according to some Islamic traditions, will rid the Earth of evil prior to the Day of Judgment. In recent times, the practice of incorporating a white *shahada* (literally 'testimony': a creed that forms one of the Five Pillars of Islam) into the design of the flag as a military ensign – a tradition originally adopted by the eighteenth-century Hotak dynasty of the Afghan Ghilji Pashtuns – has been taken up by a number of Islamic jihadist groups including the Taliban, al-Qaeda, al-Shebaab and Isis.

the one hand and Osama bin Laden and Abu Bakr al-Baghdadi on the other there exists an optimal level of categorization that depends very much on what we are trying to categorize. Moreover, when it comes to influencing others – when it comes not to how we *generate* categories to define reality *ourselves* but rather to how we *utilize* categories to define reality for *others* – the same rule applies. And, with apologies again to Freddie, less is nearly always more.

In fact, as we spool back in evolutionary time in search of the prehistoric origins of black-and-white thinking we find that when it comes to shaping opinion, when it comes to changing minds and getting others on side, *three* is the magic number. Deep in the mists of our dark, Darwinian past we unmask a secret golden triad of dyadic ancestral super-categories that still, to this day, exert such a profound and powerful influence over our ever-impressionable brains that when they're strategically invoked by those who seek to persuade us never fail to bring us on side: *Fight versus Flight, Us versus Them* and *Right versus Wrong*. In the hands of a Bin Laden or a Hitler such categories can cost millions of lives. But when judiciously applied in the service of common good they are capable of influence miracles.

An example: 62,000 people attended the dress rehearsal of the opening ceremony for the London 2012 Olympics but only a handful of those present leaked the contents of the show on social media. That's pretty amazing when you think about it – especially in light of the pressures, expectations and temptations associated with modern-day download culture. Speculation abounds as to why this might've been the case but one theory is particularly intriguing. On the evening of the extravaganza the London 2012 artistic director, Danny Boyle, addressed the lucky attendees assembled before him in a hushed Olympic Stadium at Stratford's Olympic Park and asked them not to 'keep it a secret' but rather to 'save the surprise'. A trivial, insignificant detail? On first impressions, yes. But on closer inspection – if true – a flash of persuasion genius.

Here's how it works, the hidden evolutionary small print beneath the banner persuasion headlines, based on the three ancestral super-categories:

Fight versus Flight – Resist the temptation to spill the beans. Turn a deaf ear on the whisperings of human nature. We all like sharing secrets but no one likes spoiling a surprise.

Us versus Them – Let's keep this to ourselves until the big reveal, eh? We're privileged insiders. We don't want to admit outsiders to the club too early, do we?

Right versus Wrong – How would you feel knowing that something you said not only ruined the big day itself but also all the effort and hard work that went into it?

Just that simple change of word – 'secret' to 'surprise' – just a simple shift in category – 'share' to 'spoil' – made all the difference.

In fact language, we shall discover, is pretty much the key to activating any kind of category, not just these three irresistible, weapons-grade evolutionary super-categories. Because if we didn't have language, the truth is we wouldn't have anything. The function of language, we shall learn, is actually extremely basic. It is, fundamentally, to differentiate 'this' from 'that'. To affix labels to the other once it has been newly othered.

And where language ends, persuasion takes up the reins. If the function of language is to differentiate 'this' from 'that', then the function of influence is equally clear and straightforward. It is, quite simply, to make *my* 'this' *your* 'that'.

My 'spoiled surprise' your 'shared secret'.

It's a move that all of us need to master if we want to get on in life, and one directly descended from our brain's primeval predilection to see things in black and white. But there are conditions to be satisfied. Rules to be obeyed. We need to know how it works, why it works, and when it works. And, most importantly of all perhaps, how to be on the lookout when others try to use it on us.

As our journey to the heart of influence continues, we discover that just as the brain creates rough, crude, primitive maps of reality, it is *itself* a map. An antiquated, out-of-date, weather-beaten map that folds along three major creases – *Fight versus Flight, Us versus Them, Right versus Wrong* – and that knowledge of these creases and how the map opens and closes renders it a hell of a lot easier to navigate than a map that just opens at random.

We learn that the brain's default mode of 'either/or' binary categorization of the world cuts across all aspects of our lives, applies to every judgment or decision we make, and marks the standout cognitive legacy of our early forebears' savage and heroic struggle to beat the odds and survive the cutthroat killing fields of prehistory. Drawing on insights from everything from forensic science to social and cognitive neuroscience, and from intergroup dynamics right up to the hotly contested shadowlands that huddle between the frontiers of language, attention and thought, *Black-and-White Thinking* presents a unique and ground-breaking synthesis of cognitive, evolutionary and persuasion science.

In short, it offers a brand new theory of social influence. A theory of what I call *supersuasion*. A theory that has undergone considerable conceptual refinement within the cultural and political blast furnaces of four of the most dominant global news stories of modern times.

Brexit. Trump. Coronavirus. And the rise of Islamic fundamentalism.

Everything 'out there' sits along a continuum. Grade points. Skin tone. Even, despite the binary ballot of the jam jars, our preference for cats or dogs. Everything out there is grey. But in order to make sense of reality, in order for us to work out how its multitude of different elements relate to and interact with each other, we need to be able to dissect the amorphous, unstructured continua into smaller, sharper, self-contained bite-size sections. We need to draw lines in the world's interminable greyness to create an illusory chequerboard surface upon which we can move, sense and reason like rational, thinking chess pieces in an orderly, predictable and rule-based fashion. We need, as the mathematician and philosopher Alfred North Whitehead once put it, to create a 'fallacy of misplaced concreteness'.

Chess works because the board is black and white. Life works because our *brains* are black and white. But wisdom lies in knowledge of the grey; in the deeper understanding that although, as cognitive grandmasters, we are destined to play the game, the squares on the board, indeed the very board itself, do not exist.

Etched in black and white is the oldest, simplest, most powerful truth there is. The truth that Freddie missed. Fantasy *is* real life. Because *everything* really matters. Everything really matters to me.

CHAPTER 1

The Categorization Instinct

: :

Progress is man's ability to complicate simplicity.

THOR HEYERDAHL

WHEN LYNN KIMSEY turned up for work one balmy summer's morning back in 2003 she had no idea that the events of the day would transform the next four years of her life into a sinister, serpentine subplot of a macabre psychological thriller. That evening, prepped, briefed, and working for the Attorney General, she would be riding the freeway home as the real-life embodiment of Dr Pilcher, the cross-eyed 'bug guy' from the film *Silence of the Lambs*.

Earlier that year, on the morning of Sunday 6 July, a woman called Joanie Harper, her three children and her mother, Earnestine Harper, had all attended a small neighbourhood church service in Bakersfield, California. It was a big day for the family. For Joanie's youngest child, six-week-old Marshall, it was his first time at church. After the service the family went out to have lunch at a local diner. They then headed home for an afternoon siesta before going back to the church for the evening service. Joanie and her children all slept in the rear bedroom, while her mother slept in another bedroom at the other end of the house.

That had been the plan, anyway. But that evening, no one saw the Harpers at the service.

On Tuesday morning a family friend, Kelsey Spann, decided to look in on Joanie, her mother and the kids. Not only had they not shown up at church on Sunday evening, no one had seen or heard from them since. And they weren't answering calls. Maybe something was wrong.

Kelsey went to a side door to let herself in to the house with a key given to her by Joanie for safekeeping. But the door wouldn't budge. The key turned in the lock but there seemed to be something on the other side preventing it from opening. She walked around to the back of the house and tried the sliding glass door. To her surprise, it opened. That was extremely odd. Joanie was always checking that door to see if it was locked. Kelsey entered the house and made her way to Joanie's bedroom.

At 7a.m. that Tuesday morning a 911 call came through to Bakersfield police. It was placed from 901 3rd Street. Joanie's address. The scene that awaited them when they arrived shocked even the most experienced law enforcement officers.

Joanie was found lying face down on the bed. She had been shot three times in the head with a .22 calibre pistol and twice in the arm. She had also been stabbed seven times.

Four-year-old Marques Harper was also found on the bed, eyes wide open. He had a gunshot wound on the right side of his head and the fingertips of his right hand had been bitten to the bone. Investigators concluded that this was a fear reaction. Marques must have seen the killer and had instinctively stuck his fingers in his mouth.

Lyndsey Harper, aged just two, was found at the foot of the bed still wearing her little blue dress from church. She had been killed by a single gunshot wound to the back.

Earnestine, Joanie's mother, was found in the hallway with two gaping bullet holes in her face. She had been shot at close range. By her side lay a pistol. Whoever the intruder might have been she had clearly intended to go down fighting.

Finally Marshall, Joanie's six-week-old son, initially thought to be missing, was found lying next to his mother concealed under a pillow. Just like his sister Lyndsey, he had also been killed by a single gunshot to the back.

The police investigation into the murders quickly gathered pace and it wasn't long before a prime suspect emerged. Forty-one-year-old Vincent Brothers, Joanie Harper's estranged husband, was a pillar of the Bakersfield community. A family man, Brothers had a bachelor's degree from Norfolk State University and a master's degree in education from California State University, and, having initially joined a local elementary school in 1987, had, over a period of eight years, risen through the ranks to become its vice-principal.

But Brothers had a dark side. Though Joanie had certainly loved him, and had tried hard to keep things between them on an even keel, their relationship had been on and off. It had been off in 2000, barely a month after they married. Yet later that year Joanie had given birth to Lyndsey, their second child. Just as with their first, Marques, born a couple of years earlier, Brothers wasn't present at the delivery. In 2001 the marriage was annulled, with Brothers citing irreconcilable differences and Joanie citing fraud. Allegedly, at the time of the marriage, she'd been unaware of Brothers' two previous wives.

That knowledge, in hindsight, could have saved her life. In 1988, Brothers had been sentenced to six days in jail for abusing his first wife and was put on probation. In 1992 he had married again, only for his second wife to sue for divorce the following year claiming that he was violent and that he'd threatened to kill her. Then, at his home in 1996, Brothers had sexually harassed a female employee of the school where, only the previous year, he'd taken over as vice-principal. According to district records, the woman claimed that Brothers had dragged her into his bedroom and had beaten her and taken pictures of her. Though she had reported the incident to the local authorities the police had dissuaded her from filing charges on the understanding that Brothers was 'a role model in the community'.

In January 2003, in Las Vegas, Joanie and Brothers had got married a second time. But in April Brothers once again moved out of the house due to friction between him and Joanie's mother, Earnestine. In May, baby Marshall was born. Now, six weeks later, he was dead. Clearly, the prosecution argued, when the trial finally opened to a blaze of publicity in February 2007, here was a relationship that was demonstrably volatile and a man who was not only violent but adulterous. In fact, it was Brothers' alleged string of extramarital affairs that formed the central thrust of the prosecution's case against him. The primary motive for the killings, it was suggested, was avarice: Brothers had wanted to relieve himself of the financial burden of supporting his growing family.

He was arrested in April 2004. The charges were five counts of first-degree murder. At the trial he pleaded not guilty.

Brothers' alibi was geography. At the time of the murders, his defence claimed, he was actually on vacation some two thousand miles away in Columbus, Ohio, visiting his brother Melvin, a brother, incidentally, that he hadn't seen for ten years. There was a rental car agreement – for a Dodge Neon, later seized by detectives – to prove it, plus a pair of credit card receipts for items purchased in a store in North Carolina on the day that the murders had taken place. Indeed, it was to his mother's house in North Carolina that police had first tracked Brothers to break the news of the horrific slayings.

But gradually things started to unravel. A more detailed examination of the credit card receipts, combined with analysis of security camera footage taken from the store at the time when the items were bought, revealed that it was, in fact, Melvin who had made the purchases, having appropriated his brother's card and forged his signature.

Moreover, closer investigation of the rental car confirmed that while Brothers had indeed leased the Dodge in Ohio he'd also put in excess of 5,400 miles on the clock. Unlikely though it was that such a journey would, under normal circumstances, have been completed within three days, that, the prosecution contended, was more than enough distance for him to have driven to Bakersfield and back.

Yet the evidence, though compelling, was still only circumstantial. Brothers may have been an adulterer, countered the defence, but that didn't make him a murderer. If it did then, statistically, at least one third of the jury would be up for trial themselves. Furthermore, the fact that Melvin had used his brother's card in a store in North Carolina in no way implied that that brother was on the other side of the country gunning down his family. He may, instead, have been waiting outside in the parking lot. Nor, for that matter, did the 5,400 miles chalked up in the Dodge necessarily entail that Brothers had driven to California. That total, after all, might in reality have been clocked up anywhere. Indeed, to reduce the argument *ad absurdum*, Brothers could in theory have run it up without even crossing the Ohio state line.

What was needed was fact. Not inferences or assumptions or guesswork. But solid, irrefutable proof.

Things started to look up for the authorities when a neighbour in Bakersfield reported seeing Brothers in the vicinity of the Harper home around the time of the murders. Then again, they could have been mistaken. Was it Brothers or wasn't it? Could they swear that it was him? The case was proving exasperating. All the evidence consistently pointed at one man but there was no forensic slam dunk to put the case to bed.

The key had to lie with the rental car. Those miles. There had to be a way of tying Brothers, the murders and the tarmac together.

But what was it?

Nature's bugging device

On 25 July 2003, two FBI agents and a Bakersfield police officer strolled through the doors of the Bohart Museum of Entomology at University of California, Davis with a car radiator in their hands. The grille was spattered with bugs and they wanted to know what they were. Not that there was anything unusual about the bugs, at least not in central California. But whether or not something is unusual is, by definition, wholly dependent on context. The bugs may not have been uncommon in California. But what about other places, the officials wanted to know? Ohio, for instance? Or North Carolina?

A slight, studious-looking woman with sharp hazel eyes and a hardy, no-nonsense pixie cut greeted the trio. In her mid-forties, Lynn Kimsey was professor of entomology at UC Davis and the museum curator. With a specific interest in the biogeography of insects, in particular Californian ones, there was no one better qualified to answer the lawmen's questions. If that radiator had been anywhere west of the Rockies Kimsey would be able to tell. She could read those bugs like an ecological X-ray. She examined the grille and took it away for analysis.

Last year, I dropped in on Kimsey in Davis to chat about the case. Almost two decades on it was still fresh in her mind. 'At the time I had no idea that I was actually taking part in a murder inquiry,' she told me as we wandered around the Linnaean labyrinth of meticulously labelled insect trays deep in the bowels of the museum. 'They left that bit out, and probably for good reason. Many scientific investigations are best performed blind, without the researcher knowing what particular puzzle they're trying to solve. That way you don't risk getting in the way of what you're trying to do, contaminating the scientific method with your own expectations. It's very subtle. You can do it without realizing. And as an expert witness in a murder trial that can have serious consequences.'

My eyes trail across the immaculate alignment of white secretarial tags slotted on to the front of the shallow sycamore drawers – Cornell drawers, Kimsey explains. *Lepidoptera*: butterflies and moths. *Orthoptera*: crickets and grasshoppers. *Hymenoptera*: bees, ants and wasps. No other creature on the face of the earth is more punctiliously preserved in death than the insect.

We pause by a tray labelled '*Xanthippus corallipes pantherinus*'. Kimsey slides it open.

'As it turned out we found thirty individual insects on various portions of the radiator,' she says. 'Or rather, parts of insects: wings; legs; an abdomen; sections of the abdomen; a head and abdomen with no wings or legs . . . But when it came down to it, it was six that told the story.

'For a start, there were two beetles that we know only live in the eastern United States. Then there were two true bugs, *Neacoryphus rubicollis* and *Piesma brachiale* or *ceramicum*, which are only found in Arizona, Utah and Southern California. They were on the air filter. There was a large golden paper wasp, *Polistes aurifer*, minus a few wings and legs – that's found mostly in California but has been sighted as far east as Kansas. And then there was this little fella, *Xanthippus corallipes pantherinus*, more commonly known as the red-shanked grasshopper. Or rather, what was left of him. We identified him from one of his back legs. The inside portion, the shanks, are bright red.'

Kimsey takes out the tray and hands it to me. I peer down through the glass lid. It doesn't require too great a leap of the imagination to figure out the thinking behind the name. The legs are indeed bright red, glowing like embers beneath the mottled grey ash of the body. I slip the tray back into its drawer and slap it shut. It was incredible to think that a simple bug like that had the power to put a man on Death Row.

'And where does *Xanth* . . . the red-shanked grasshopper hail from?' I ask, as we retrace our steps back to *Lepidoptera*, the moths and butterflies quarter.

Kimsey smiles. '*Xanthippus corallipes pantherinus* is found no far-ther east than Kansas and central Texas,' she says. 'So, all things considered, yes . . . the car that that radiator belonged to had at some point been in the eastern United States. But equally, at some other point, it also had to have passed through states west of Colorado which, as it turned out, was consistent with the hypothesis that Brothers had driven west from Ohio on either Interstate 70 or 40.'

The lawmen were more than satisfied. When, a week or so later, they pitched up back in reception to pick up the radiator and hear what Kimsey's categorical skills had turned up, what she had to tell them was music to the ears. Her entomological satnav was as good as the real thing. It was as if Brothers' rented Dodge had been fit-ted with its very own mobile tracking device that charted his progress every few hundred kilometres.

It was the final nail in his coffin. Or, more aptly perhaps, the final pin through his wings. Kimsey duly gave evidence in the Bak-ersfield Superior Court and, on 15 May 2007, the jury convicted Brothers of the murder of his wife, his three children and his mother-in-law.

We stop by another drawer. The label on the front says *Acheron-tia styx*, the Death's-head hawkmoth from *Silence of the Lambs*. Kimsey opens it and hands me the tray. 'The judge rejected the option of life imprisonment without parole and sentenced Brothers to death,' she states, matter-of-factly. 'Today he's in San Quentin awaiting execution.'

The briefest of shivers runs through me.

'How do you feel about that?' I ask, studying the contents. 'It was your evidence that put him there.'

She shrugs. 'I don't feel anything,' she says. 'I mean, he put him-self there by what he did. I was just doing my job. Doing what I do every day. Sorting things into boxes.'

Putting the cat into categorization

In 2005, the year after Brothers was arrested, the American developmental psychologist Lisa Oakes conducted a study at the University of Iowa that shone a fascinating light on how we all, to use Kimsey's phrase, sort things into boxes. Oakes was interested in how early this box-sorting, or, as I like to refer to it, 'categorization instinct' kicked in. Was it something that the brain just *did*, like hearing, or smelling, or crying? Or did we somehow need to learn it?

To find out, Oakes took a bunch of four-month-old babies and flashed pictures of cats at them on two computer screens side-by-side in her lab. The cats were presented simultaneously, and in pairs, one on the left-hand screen and one on the right. For the fifteen-second duration that each pair were on the screens an observer recorded how long the babies spent looking at each cat, the orientation of infant attention to a stimulus representing a standard measure of its novelty.

But then came the catch. After the infants had attended to six pairs of cats and had started becoming familiar with them – as indicated by a decrease in looking times over the course of the six trials – Oakes snuck in either a new cat that they hadn't seen before, or a dog.

The rationale was simple. If the babies looked longer at the dog than at the new cat then that would suggest they saw the dog as being more different to the familiar cats than the new cat. In other words, it would demonstrate that their brains were processing dogs in a different way to the way they processed cats, ascribing them to a new category. If, on the other hand, the babies' attention span did not selectively increase on seeing pictures of dogs then that would imply their brains were treating dogs and cats as one unitary, inclusive category. That of 'animal'.

What Oakes found was extraordinary. Despite the fact that the infants had had such minimal prior exposure to the dogs and cats; despite the fact that, at just four months old, they had yet to acquire the words for 'dog' and 'cat'; despite the fact that dogs and cats are, when you think about it, actually pretty similar to each other – both have four legs, two eyes, fur and a tail – they looked longer at the pictures of the dogs than at the pictures of the novel cats. The brain, at just four months, is already sorting the outside world into boxes.

By extraordinary coincidence – and a modicum of forward planning – a few hundred metres across campus from the Bohart Museum of Entomology lies the UC Davis Center for Mind and Brain, where Oakes, having left the University of Iowa back in 2006, now heads up the Department of Psychology's Infant Cognition Lab.

We talk.

I tell her about my meeting with Professor Kimsey and how entomological taxonomy had snared a multiple murderer. She's impressed. Not that that kind of minutiae ever darkens the computer screens in her lab. Cat and dog are about as nuanced as it gets.

The world, she explains, is a complicated place. When we first enter it, it appears, as the father of Western psychology William James once put it, as a 'blooming, buzzing confusion'. It's a problem that needs a solution. As with most problems, it is easier to deal with once it's been 'cleaned up'. And so our brains begin sorting the blizzard of incoming data into separate, more manageable piles. Eyes, noses and mouths become faces. Things that bark, neigh or moo and that have four legs and a tail become animals.

'Just imagine what the world would be like if our brains couldn't form categories,' Oakes says. 'Even the simplest things that we take for granted on a daily basis would pose a huge challenge. You walk into your friend's garden and they have a new sprinkler system. But you don't have the category "watering device". "What's that object over there in the middle of the lawn?" you think to yourself. "I don't recognize it. Is it dangerous? Is it something that could kill me?"'

If we didn't have the ability to categorize, Oakes continues, waking up each morning would be like getting out of bed on a new planet. Hairdryer: what's that? Is it trying to attack me? Television: who are those people in there? Are they trying to talk to me? Washing machine: hmmh . . . do I put my head in there?

Categories enable us to navigate the world, object by object, person by person, in a predictable and orderly way so that our journey through life doesn't just consist of an endless string of random, novel and essentially meaningless interactions but is planned, controlled and purposeful.

In that sense babies might be considered the R&D arm of the species, I suggest?

'Absolutely,' Oakes says. 'Initially, when we are little, we perceive the world in expansive, broad-brush categories. Like "plants" and "animals", for instance. Then gradually, over time, as we hone our categorization skills, these categories become more delineated. We see flowers and trees. Dogs and cats. Birds and fish. Large and small. Cuddly and not-so-cuddly. And then, as experience and development continue, we make ever more fine-grained distinctions. We separate Chihuahuas from Labradors; Persians from Siamese; deciduous trees and evergreens; small red cardinals and large pink flamingos; sharks and dolphins.'

Further down the road, she tells me, we become even more picky still. We see red pines, white pines, acacia trees and orchids; golden eagles, grey geese, robins and sparrows; red admirals, purple emperors, orange tips and meadow browns. Eventually, in adulthood, if we enter the fields of botany or biology, our taxonomic systems become so fine-tuned that, much to the irritation of the friends who are out walking with us, we find it impossible not to lapse into unintelligible jargon when they point at the pretty flower.

Or, if you enter the field of entomology, much to the irritation of multiple murderers, you find it impossible not to break into incomprehensible Latin when the police show up on your doorstep with a radiator full of dismembered bugs and moths.

Irrespective of your classification credentials, the punch lines are identical. Predictability, expectation and the minimization of uncertainty. The exact same principle that applies to four-month-old babies on the nursery slopes of categorization also applies to serious Alpine categorizers like Lynn Kimsey high above the taxonomic snowline. Categorization, as Oakes elucidates, is about orderly and efficient navigation, no matter where or when it is done.

Which raises, for the rest of us, a fundamental question: what level of categorization is considered optimal for greatest efficiency within the course of our everyday lives? If our categorization instinct evolved to reduce complexity then doesn't categorizing the ordinary and commonplace – items like dogs or houses, for example – with the taxonomic voracity of a forensic entomologist somehow defeat the object?

Nine thousand miles west, across a cold blue sea and a red-hot continent, it's a matter I raise with Professor Mike Anderson, dean of psychology and exercise science at Murdoch University in Perth, Western Australia.* Anderson, a Scot, is one of the world's leading experts in categorical perception. In particular, how it develops in children.

We categorize the world at three different levels, he explains. Superordinate, basic and subordinate. It means that when we pigeonhole something we can be as general or as specific as we like.

---

* After a long and illustrious career in the field of cognitive psychology Mike has since retired from academia.

'It's not a perfect metaphor but think of it in terms of a family tree, with the more general, or superordinate, classifications being at the top and the more specific, or subordinate, classifications being at the bottom,' he suggests. The superordinate level classifications are like the parents, the basic level classifications are the children, and the subordinate level classifications are the grandchildren and great-grandchildren.

'Here's an example,' Anderson continues. 'Imagine I were giving you directions and I told you to turn right at the end of the road by a square concrete structure with a door, four windows and a driveway that had a mammal which barked, had four legs, fur and a wagging tail in the garden. You'd think I was a little bit strange! That's a lot of words. Why didn't I just tell you to turn left at the house with the dog? Because as soon as I say the words "house" and "dog" your brain will automatically fill in all the other details.

'On the other hand, imagine I said: "Take a right at the vermiculated artisans dwelling with a mansard roof and a Bergamasco Shepherd out front." Again, you'd think I was nuts. But this time, instead of being overly general I'm being overly specific. Unless you were an architect with an interest in rare dog breeds you'd still be none the wiser where to go. And even if you *were* an architect with an interest in rare dog breeds it would sound rather odd.'

So we go for the happy medium. In general conversation we choose what we call basic level categories to convey, acquire and organize information because they are the ones that save us most time and energy and enable us to communicate most effectively: the reason we evolved the ability to categorize in the first place.

Broadly speaking, Anderson elucidates, for everyday life these basic level categories are the most optimal. And, he points out, have a so-called 'privileged' status. In other words, if they were taxonomic stars in the boundless category firmament they would be the ones most visible to the naked eye; they would be the ones shining brighter than the rest; the ones that, if you were plotting your course by them, you'd be best advised to follow.

'Ask a four-year-old, for instance, whether a baby cow would still grow up to moo and not oink if it was raised in a family of pigs and they'll say yes,' Anderson tells me. 'Even at four kids recognize that a baby animal will grow up to acquire the features of other animals in its category regardless of where it is raised. In the same vein, ask a five-year-old whether a porcupine transformed in such a way as to be outwardly indistinguishable from a cactus would still be a porcupine, and they'll say yes. Despite what it might look like on the outside, it will still be a porcupine.'

From a very young age, then, kids 'privilege' basic category distinctions – cow, pig, porcupine – over similarity in appearance when making inferences about what characteristics different animals might have in common. Not only that, but they also seem to understand that basic level category membership is fixed; that cows are cows and will still moo irrespective of whether they grow up in a family of pigs, or llamas or wildebeest.

'So yes, in answer to your question,' concludes Anderson, 'in everyday life it's the basic level of categorization that is both the most useful and the most natural. But it depends on what you mean by everyday life. Everyday life can mean a lot of different things to a lot of different people. For example, there's a ton of research showing that the more we know about something, the more likely we are to use finer levels of detail when we categorize it. Experts in a particular field will categorize it to death. That means that what *they* might consider optimal when they talk amongst themselves might not be optimal to outsiders.'*

---

* There is evidence to suggest that such expertise may also extend to how well we know ourselves. The American cognitive psychologist Lisa Feldman Barrett has coined the term 'emotional granularity' to refer to striking individual differences that exist in our ability to put feelings into words: to categorize 'primary'

Take biology as a case in point. Within the system of biological taxonomy there are seven levels of categorization: kingdom, phylum, class, order, family, genus and species. For biologists, the optimal level is genus – which comes from the Latin word for 'kind' or 'type' – the level *above* the one that most people would probably find handiest, species.

'Going back to dogs then,' Anderson continues, 'the word "dog" is actually a species-level descriptor – *Canis familiaris* – referring to a member of the genus *Canis*. Other species-level members would include *Canis lupus*, the wolf, or *Canis latrans*, the coyote. So when we talk about there being an optimal level of categorization we do have to exercise a little bit of caution. It really does depend on the context: on the bigger picture, the wider situation and whatever is being categorized.'

Which is why, in Davis, one balmy summer's morning back in 2003, the FBI rolled into the Bohart Museum of Entomology and handed the beat-up radiator of an impounded Ford Dodge to taxonomist extraordinaire Lynn Kimsey. Exceptional circumstances called for exceptional powers of categorization. And the curious case of the one-legged red-shanked grasshopper would've taxed even Sherlock Holmes.

---

or 'basic' emotions such as anger, fear and happiness. Those low in emotional granularity will typically use words such as 'angry', 'sad', or 'afraid' when describing their reactions to upsetting or unpleasant events, and words such as 'happy', 'excited', or 'calm' to capture positive feelings. In other words, they are less emo-diverse. In contrast, individuals who are high in emotional granularity employ a far richer and more nuanced vocabulary when articulating how they feel – they categorize their emotions at the subordinate, as opposed to the basic or super-ordinate level – drawing on primary emotion sub-categories such as shame, guilt and regret.

CHAPTER 2

A Heap of Trouble

❧

*The continuum is that which is divisible into indivisibles that
are infinitely divisible.*

ARISTOTLE

IN OCTOBER 2004, Paul Sinton-Hewitt wasn't feeling great. He'd
just got the bullet from a well-paid marketing job and was
injured. Not from being laid off – no real bullets had been fired in
his dismissal – but from pounding the streets of his West London
neighbourhood in readiness for the London marathon. He'd been
to the physio and the physio had shaken his head. Knees don't clear
up in a hurry. If anything, they deteriorate further. As always in
life, the timing just couldn't have been worse. He withdrew from
the race, adjourned to the pub and had a bit of a mull. Out of a job
and now out of the marathon, he felt crushed. Like an eviscerated
bug in the radiator grille of life.

Fifteen years on and Paul and I are sitting in that same pub. It's
just down the road from the running club in Richmond of which
we're both members. I return from the bar with some drinks. We
take a few sips and look around.

'I had a choice,' Paul tells me matter-of-factly. He is warm, wiry,
grey and softly spoken. 'I could've moped around the house feeling

sorry for myself and been a victim. Or I could've got off my arse and taken the opportunity to *do* something. To give something back. To make a difference to other people's lives while I worked out what to do with my own.'

To his considerable credit – and the immeasurable good fortune of millions of fellow runners around the world – he chose to do the latter and on 2 October 2004, thirteen lycra-clad pioneers, unwitting revolutionaries, converged on Bushy Park, south-west London, at precisely 8.45a.m. on that chilly Saturday morning to walk, jog or go full pelt – it was entirely up to them how they went about it – over a distance of 5 kilometres. Afterwards, in a nearby café, Paul typed up the results while the athletes, none of whom were going to give Mo Farah much to worry about, went for a post-race fry-up.

'Initially it was just me and a group of friends,' Paul explains. 'I organized a run around the local park for them while I hung about with a stopwatch. I wouldn't say that I had a vision back then. Not as such. But in the back of my mind I did want to do something that was fun, socially inclusive and, above all, free – something that would encourage individuals of all ages and physical abilities to take regular exercise, embrace a healthier, more active lifestyle and, most importantly of all, stick with it.'

Sixteen years, 715 locations, 166,896 events, 34,853,835 individual runs and 174,269,175 kilometres later,* the bright and early Saturday morning Parkrun has now become for many – from the terminally ill and those recovering from serious illness and injury to celebrities and Olympic gold medallists – a staple of both their weekly social calendars and their wellness and fitness programmes. Not to mention a national and international phenomenon. From Australia to Japan. From Singapore to Eswatini.

Parkrun began in the same year as Facebook, but it's fair to say from

---

* These statistics are correct at the time of writing and apply to Parkrun UK only. But Parkrun is organic and adds to the tally with every week that passes.

the moment they both started out that Mark Zuckerberg and Paul Sinton-Hewitt were never going to end up next-door neighbours.

'A lot of people make a lot of money out of running and sometimes it's completely uncalled for,' says Paul. 'I wanted to change all that; throw a spanner in the works; become – to use one of those trendy corporate buzzwords – a disruptor. Everyone has a right to do what they want. Especially run! So I thought: why charge them? Why make them pay for something so simple and natural?'

But in 2017 the parish council of Stoke Gifford, a dormitory village in the northern suburbs of Bristol, south-west England, elected to break with Parkrun tradition and resolved to do just that: charge people to run. The three-year-old event held in nearby Little Stoke had, according to the chairman of the council, Ernest Brown, become a victim of its own success. It had started modestly enough with just a few dozen runners, but had snowballed into a weekly jamboree comprising several hundred entries.

Three hundred feet 'pounding the paths' every Saturday morning, Brown announced at the time, had caused 'extra wear', a growing concern that had apparently left fellow incumbents of the council with little or no recourse but to request that those participating contribute 'a small monetary amount towards the upkeep'.

At Parkrun HQ alarm bells started to ring. Brown was unleashing a dangerous and unnecessary precedent – one that threatened to undermine the very founding principle of the enterprise. The cost of entry was only £1. But that was hardly the issue. It wasn't just *beside* the point, it was in a totally different universe to it. There was uproar. And the event, much to Paul's disappointment, was shelved.

'The whole idea of Parkrun, which the council just didn't seem to get, was that the events were put on free of charge,' he tells me, still clearly bemused by the decision. 'That was non-negotiable. I mean, think about it. If one has to pay while the other five hundred dotted around the world do not, then there's no way of telling where it would all lead.'

Well there is, I point out, respectfully. Tins would start to rattle all over the place.

Other arguments and counter-arguments abounded. The extra wear, a bunch of physics boffins concluded, was negligible. If the increased footfall every Saturday morning compressed the asphalt of Little Stoke by something in the region of $1mm^{-20}$ per race then that would entail a total drop in elevation in the order of magnitude of a Rizla Super Thin cigarette paper by the time of the next Ice Age. Surely, it was a small price to pay for the immeasurable enhancement of the participants' subjective quality of life as well as the objective health benefits that would help to conserve NHS resources? Even former women's marathon world-record-holder Paula Radcliffe was drawn into the fray, describing the council's decision as 'short-sighted'.

We grab some menus and move into the restaurant area of the pub for a bite to eat. As we sit down I drop a bit of a bombshell. The real difficulty, I suggest to Paul, appeals more to the principles of *meta*physics than it does to physics. Where does one draw the line? At what point, precisely, do the numbers get too high? What is the threshold beyond which the run metastasizes from a casual, meet-you-by-the-duck-pond congregation of knobbly-kneed diehards to an *event*, a full-on stampede of jump-suited Elvises, iridescent Tele-tubbies and the obligatory slew of dinosaurs, superheroes and brick-laden Royal Marines?

Back in the day, Paul's original band of thirteen co-runners would have been deemed perfectly acceptable by the great and the good of the parish of Stoke Gifford. Equally, a 400-strong legion would not. That glaring disparity is easy to capture from a distance, perceived down the barrel of a wide-angle, three-year lens. But the closer one studies the Parkrun transformation, the more the pixels begin to lose their sharpness and the softer and fuzzier the logic then becomes. The difference between 13 and 400 is easy to get. Fifty and 350? Sure. But what about the difference between 175 and 225? Or between

195 and 205? What about the difference between 199 and 201? If, as we saw in the introduction, we need to press creases in reality to make sense of the world around us – turn grey into black and white – then how, precisely, do we know where to run our irons?

Paul tells me there *is* now a cut-off. It's 300. But that's driven solely by the runners themselves and not by the local authorities. From the feedback they receive, when the numbers start getting around the 300 mark, people tend to find it too congested. Which is precisely my point. *Around* the 300 *mark*. But if you had 301 runners in the park would people notice the difference? What about 302? Or 303? Paul shrugs. He gets it. But at the same time observes that there has to be a cut-off somewhere.

I get *that*, too. A line certainly needs to be drawn. But the nearer one stands to the decision-making canvas, the more one zooms in on the individual brush strokes of categorical reasoning, the greater the reduction in analytical clarity. The more the picture degrades and falls apart.

It's not a new phenomenon, of course. The conundrum has form. Several thousand years ago, for example, an Old Testament Bible story in the book of Genesis finds God Himself in a strikingly similar position to that of the esteemed elders of the parish of Stoke Gifford. The townsfolk of Sodom and Gomorrah had been engaging in certain practices that perhaps, in hindsight, were never going to secure them the seal of divine approbation and the Good Shepherd was cross. So cross, in fact, that He had decided to teach the poor idolatrous unfortunates the error of their ways and firebomb them out of existence.

Abraham – more empathic and emotionally intelligent than his boss – expresses serious reservations over the planned intervention and remonstrates at length with his apocalyptically combustible superior over the wisdom of His incendiary intentions, placing particular emphasis on the potentially negative PR that might ensue from instigating a genocidal fireball.

But if one sets oneself up to make these kinds of judgments then one has to have the stomach to wield the brimstone *somewhere*. One has to be prepared to draw the line. But where? Where does one draw it with any degree of assurance?

Scripture is illuminating here. Instead of destroying the entire city – the 'godly' along with the 'wicked' – Abraham, according to the Genesis account, manages to convince God to 'do what is right', to cut some celestial slack to the infidels of Sodom and Gomorrah just so long as there prevails within their dissolute, degenerate ranks a certain number of innocent and virtuous souls. The number begins at 50 and is gradually whittled down, upon Abraham's indefatigable haggling, from 45 to 40 to 30 to 20 to 10. But at the end of the day (a particularly salient cliché in this case) 10 was still an arbitrary cut-off . . . or may, alternatively, simply have been the point at which Abraham decided to fold.

So where *does* one draw the line? The answer, quite simply, is: you can't. Every cast of the black-and-white bones is a loser. Every throw of the line-drawing dice is a dud. Which means that we have a problem on our hands; a slapstick philosophical banana skin that is dangerously, reason-splittingly slippery. Are we really going to turn someone away from Parkrun purely on the grounds that their last-minute entry into the event is going to tip numbers over the fee-paying threshold and necessitate, as a consequence, that all of their fellow participants suddenly have to cough up £1 for the pleasure of their company? Are we really going to visit a cataclysmic fusillade of Jehovah-grade napalm on two civilizations just because one of the good guys leaves town and diminishes the ranks of the chosen to an incendiary sub-quorum of nine?

A little more recently, on the evening of Thursday, 12 March 2020, as the Coronavirus crisis began to deepen, consider the Chief Advisor to the Prime Minister Dominic Cummings's purported 'Domoscene conversion' at a SAGE (Strategic Advisory Group of Experts) meeting when, having reviewed the increasingly desperate chain of events unfolding in Italy, he underwent a dramatic change of heart and desisted from advocating a strategy of so-called 'herd immunity' in favour of a policy of unprecedented social distancing.

Advocates of the herd immunity approach subscribe to the view that pandemics should be permitted to run their course in order to enable a significant (younger) proportion of the population to build up resistance, thus preventing a catastrophic 'second wave' of the virus or disease at some future point in time. Strategically managing outbreaks in this way, proponents of the policy observe, minimizes damage to the economy by allowing more people to potentially remain at work but does, admittedly, place the most vulnerable members of society – the elderly and those already suffering from underlying health conditions – at increased risk of death and serious illness.

It's important to note that Downing Street vehemently denied any claim of a herd immunity U-turn, which originally appeared in an article in the *Sunday Times*, dismissing it as 'defamatory fabrication'. But let's just put ourselves in Cummings's shoes for a moment. We're at a meeting that will change the face of British society for a generation. Imagine if we *were* faced with such a choice. At what point, if any, do the figures become justifiable?

According to a senior Whitehall figure quoted in the article, Professor Chris Whitty, the UK's chief medical officer, and Sir Patrick Vallance, the chief scientific adviser, had been advised the previous week to expect the death toll from COVID-19 to be around 100,000. Then suddenly, out of nowhere, the 'penny-drop' moment hits. The estimate is dangerously conservative.

'Unmitigated, the death number was 510,000,' the *Sunday Times* source goes on to explain. 'Mitigated we were told it was going to be 250,000. Once you see a figure of take no further action [sic] and a quarter of a million people die, the question you ask is, "What action?"'

We're all on the same page there.

But what if the figure was 50,000? Or 5,000? Or 5? Should *anyone* be 'allowed to die' in the pursuit of economic stability?

If so, once again, where do we draw the line?

Kicking up a sand storm

Watching from the stands at Manchester City's Etihad stadium back in May 2012, philosopher Dr Raj Sehgal witnesses the then Queen's Park Rangers player Joey Barton's mad minute and a half against the champions elect. Joey transforms himself into a one-man crime wave, elbowing Carlos Tevez in the face, booting Sergio Agüero in the back of the leg and then attempting to head-butt Vincent Kompany – a sequence of events which sees him dismissed from the field of play and which subsequently earns him a twelve-match ban.

Stunned, appalled and intrigued in equal measure, Raj decides to write to Joey. Joey, at the time, had been fond of quoting Nietzsche, and the press had got a hold of it. Perhaps, Raj mused, a deeper grasp of core philosophical principles might make him a better person. Or at least, in the short run, help him comprehend the difference between football and Mixed Martial Arts.

Much to Raj's surprise, Joey picks up the phone. He is interested in finding out more. Several weeks later there are regular sightings of him on the campus of Roehampton University in West London, sitting in on philosophy tutorials in the department that, a number of years earlier, Raj had founded. It was unthinkable. Raj was turning the baddest man in football, as *The Times* had once described him, into if not the wisest then arguably the most enlightened.

I first meet Raj in the basement of a recording studio in Central London, where I'm a guest on Joey's podcast, The Edge, and Raj, having relinquished his philosophical obligations at Roehampton a year or so previously, is in post as executive producer, twiddling the knobs and sliders behind the glass. He has an air of the Bollywood moneyman about him. Cool, calm, single-breasted graphite suit. Salt and pepper hair, swept back. And a freshly laundered, open-necked confidence. The topic of conversation is the dark side of talent. My specialist subject. We all hit it off immediately.

Several weeks later, in an inaugural exploration of what we decide to call immersive philosophy, we sit chatting over dinner in a restaurant off Leicester Square. Our first session is dedicated to Epicurus and it goes rather well. So well, in fact, that we ponder the wisdom of devoting all future sessions to this exceptionally clear-sighted man. As last orders approaches, Raj probes me about what else I'm working on at the moment. Is there, he wonders, a life outside of psychopaths, the enduringly popular protagonists of one of my previous books? I tell him about black-and-white thinking.

'Well, we have to draw lines somewhere,' he says. 'Otherwise there'd not only be no stopping us. There'd be no *starting* us.'

I agree, and recount the story of Lynn Kimsey and her one-legged red-shanked grasshopper. Right from the get-go the sole reason Kimsey was able to nail Vincent Brothers, I tell Raj, was because of those thin, fine, ever more nuanced lepidopterological lines that she was able to draw between esoteric species of bug, moth and fly in her professional capacity as entomological ninja. The fact that she could draw them in the first place was the beginning of Brothers' demise. Then, at the other end of the stop-start line-drawing spectrum, there was Parkrun. There had to be a cap on numbers somewhere down the road, I say. Literally as well as metaphorically. If there wasn't it would just be chaos. It wouldn't so much be a case of Parkrun as Park*over*run.

But where? That was the problem. At what point, exactly, does a non-event become an event . . . does black turn into white?

Raj looks at me as if I've just grown another head (which, given what was to come, would actually have been rather useful). My psychological ball, quite clearly, has just bounced over his immaculate philosophical fence and he has little intention of giving it back in a hurry. 'Have you ever, dear boy, in your intellectual travels,' he inquires, 'come across something called the Sorites paradox?'

The question, it turns out, is a rhetorical one. As I go to answer it, Raj bats me away. 'Don't worry, don't worry . . . evidently not. It's fine. I mean, there's no reason why you should've done, is there? What was it you said again when we first met? Oh yes, philosophy is just psychology without the funding. So yeah, what can philosophy possibly teach psychology?'

The answer, I discover, as I gorge myself on humble pie, is quite a lot. Line drawing has got serious form. And it started – where else? – in sand.

The Sorites paradox is one of the most devilishly inscrutable paradoxes ever devised. So impregnable has it proven to the onslaught of logic and reason that even now, some two and a half millennia after its inception, debate over how to resolve it still remains. The conundrum is the work of an obscure ancient Greek philosopher called Eubulides, a contemporary of Aristotle, and it gets its appellation from the Greek word *soros*, meaning heap.

Take a look at Figures 2.1a and 2.1b below. Figure 2.1a depicts a heap of sand. Figure 2.1b does not.

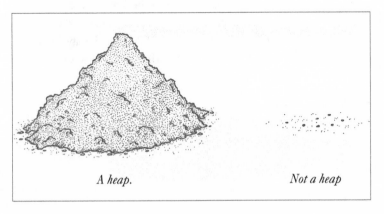

A heap.   *Not a heap*

Figure 2.1a: A heap. Figure 2.1b: Not a heap.

So far, so good. But suppose we now accept that the following two premises are true:

1. One grain of sand does not constitute a heap.
2. One additional grain of sand is too small to make a difference as to whether something is a heap or is not a heap. (In other words, adding a grain of sand to a non-heap does not make a heap.)

Suddenly we are bedevilled by the following chain of logically contiguous propositions:

1. One grain of sand does not make a heap.
2. Two grains of sand do not make a heap.
3. Nor do three, four or five grains of sand make a heap . . . and so on.

Which means (see Figure 2.2 below) that, from a purely logical perspective, none of the examples shown – including Figure 2.2d (our original heap in Figure 2.1a) – constitutes a heap because there exists, within the entire compass of both the known and unknown universe, no exact definition of the number of grains either necessary or sufficient to form one. Figures 2.2c and d may look like heaps in comparison to a and b. But there's no way of defining

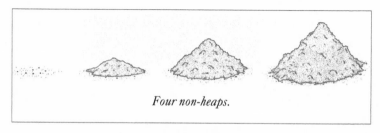

Four non-heaps.

Figures 2.2a, b, c and d: Four non-heaps.

them as such because if we start at Figure 2.2a and add to it a grain at a time then logically speaking (if the addition of one grain of sand is too small to make a difference as to whether something is a heap or a non-heap) a heap can never form.

The Sorites paradox has caused philosophers quite a few headaches down the years. Clearly, Figure 2.1a depicts a heap of sand and Figure 2.1b does not, regardless of the fast-talking maths. But it's when we start to get away from simple heaps of sand and begin venturing into the more serious business of emotionally laden judgments that the stakes, as we've already seen with Parkrun, become noticeably and appreciably higher. For grains and heaps let's substitute life and death, for instance, and enter the assisted suicide debate. There's an obvious difference between the British serial-killer GP Harold Shipman giving an unsuspecting patient a lethal injection* and a desperate, grieving husband doing the same to his pain-wracked, terminally ill wife of forty years.

Or is there? Some would say that murder is murder.

Similarly, there's an obvious difference between someone dying from a sudden drugs overdose and someone dying of lung cancer after a lifetime of smoking.

Or again, is there? One scenario might constitute a singular act of recklessness, the other an insidious, long drawn-out act of lethal self-harm. But both have the same result.

And what about a concept like gender? Facebook, as mentioned previously, has something like seventy different options for users to choose from in its dropdown menu including genderqueer, pangender and two-spirit. Even ostensibly clear-cut biological categories such as these aren't quite as simple as they seem.

Life proceeds in grains. But our attention is drawn only to heaps.

Now it is tempting to disregard the Sorites paradox as brilliant,

---

* This was the fate of at least fifteen patients in Shipman's care, for which he was found guilty in 2000, but quite possibly as many as 250.

lawyerly nonsense; as a localized philosophical oil spill off the coast of mainstream reality. But it's not that easy to dismiss and put aside. In 1834, some two and a half millennia after Eubulides began counting his grains, the German physician Ernst Heinrich Weber, considered also by many to be the founder of experimental psychology, discovered a corollary within the crystal clear territorial waters of human psychophysics. A *just noticeable difference*, or JND, refers to the minimum degree by which a stimulus must change – a tone, a hue, or a physical sensation, for example – in order for a before-and-after difference to be detectable.

Observed across all five senses, this just noticeable difference is enshrined in the following law, Weber's Law:

$$\frac{\Delta I}{I} = k$$

Where I is the intensity of the stimulus at Time 1, ΔI the threshold magnitude of increase required to render a discernible change (the JND) at Time 2, and k is the ratio between the two which remains constant for any one individual across all instances of change of the particular stimulus in question.

By way of a simple example take sound. Imagine I were to present you with a single continuous tone of 50 decibels (i.e. $I = 50$) and then slowly increase the decibel levels. Let's say that I have to increase the sound level by 5 decibels (i.e. $\Delta I = 5$) before you can detect that the volume of the tone has changed. In this case, the JND would be 5 decibels, and the ratio (k) between the initial sound level and the JND would be 5/50 = 0.1. Which means that by using this information I can then predict your JND for future sound levels too (if we started off with a tone of 70 decibels, for instance, then I would have to increase it by 7 decibels – 7/70 = 0.1 – before you could tell the difference).

The implications of Weber's Law, from both a philosophical and a biological point of view, are significant. Evolutionarily speaking, on the one hand, there exist tipping points of perception at which natural selection has decreed it important that our brains register change. Step out into the road and see a bus hurtling towards you and it's definitely going to grab your attention. Yet from the strictly logical perspective, on the other, Eubulides and Sorites still rule. We cross millions of lines every single minute. We just don't notice them. String together a sequence of consecutive *sub*-tipping point differences, and although the magnitude of change between the beginning and end points of that sequence may be noticeable, the transition from one to the other will appear changeless. Think, for example, of a clock on the wall, and the imperceptible movement of the hour hand.

The passage of time is grey. But we perceive it as black and white.

Back in 1990, the American philosopher Warren Quinn brought Eubulides and Weber face to face in a fiendishly ingenious paradox every bit the equal to Sorites: the *Paradox of the Self-Torturer*. Imagine that I have a portable device that enables me to apply an electrical current to your body in increments so tiny that you are unable to feel them. The device has 1001 settings: 0, which is 'off', and then 1 to 1,000 (excruciating pain). Now suppose that I attach you to the device with the dial initially set at 0 and offer you the following deal. You may keep the device for as long as you wish on one condition: that you turn up the dial just one single notch each week. For each turn of the dial you will receive £10,000 but you can never undo a turn and revert to an earlier setting.

Finally, here's the kicker. If, eventually, you find the pain too much to endure, you are nevertheless unable to relinquish the fortune that you have accrued and must continue to suffer the anguish for the rest of your life.

Would you take me up on the challenge?

You can see how the problem arises. 'Reality' tells you you'd have to be barking mad; that it will end, quite literally, in tears. On the other hand, however, Sorites casts its spell. No matter how many notches you ratchet up the dial, those grains of pain will never make a heap.

To this day, Raj Sehgal still feels very pleased with himself. He was right. Sorites is the problem of Parkrun and Sodom and Gomorrah, reduced, distilled and impeccably philosophized in sand. Substitute cigarettes, fudge sundaes or espresso martinis for those sub-incremental microvolts, for instance, and the plight of the smoker, the weightwatcher and the deathly, dyspeptic hangover-nurser is exposed for all to see. One more can't hurt. Can it? Often, problems arise and dangers accrue not too *fast* for us to notice, but too slow. So we need to draw lines to prevent things from going too far.

Yet as I take a few moments to get my head around it, the first thing that strikes me as I ponder the ramifications for everyday life has nothing to do with the perils of over-indulgence. With fags, or fun runs, or fudge sundaes. But rather with how we communicate with each other. With how we persuade and influence.

Some two and a half thousand years ago, in Plato's *Republic*, the Greek philosopher Socrates famously likened our sensory appreciation of reality to that of a group of troglodytes living chained up inside a cave, their backs to the entrance and their eyes looking straight ahead, surveying imperfect reflections of the outside world stream across a wall in front of them. These shadows, Socrates postulated, comprise our vague, nebulous, ill-defined representations of pure, unsullied reality: a reality composed of eternal, imperishable archetypes.

But he was wrong.

Rotate the metaphor, and the observer, 180 degrees and psychology turns the allegory on its head. In direct contrast to what Socrates thought, it is *reality* that's fuzzy around the edges, not our less than perfect perception of it, and *we* who render it 'pure' through our innate propensity to categorize the world. We see patterns in our surroundings. We caricature what's around us. We draw lines in the sand in order to compartmentalize meaning from a graduated, greyscale, infinitely continuous environment.

Which poses a bit of a problem. Not least of all when it comes to making decisions; to us making up our minds on any given subject and considering the options available. Take, for example, an issue like health and the headlines it attracts in the media. 'Sleep more, stay healthy', screams one caption. 'Forget 8 hours sleep – we just need 6', shouts another. A banner reading 'Stress takes years off your life' is pitted against one which says 'Stress can be good for you'. 'Face masks cannot stop healthy people getting Covid-19' stands alongside 'Boris Johnson says Britons SHOULD wear face masks when the country comes out of lockdown'. Polemical, conflicting, attention-grabbing headlines like these are the mainstay of the media. And with good reason. It isn't truth that grabs our brains. It is certainty. It isn't balanced, considered analysis that captures our attention. It is bold and confident assertion.

The implications for persuasion and influence are considerable. If the construction of a rational argument consists of the controlled, systematic addition of single grains of carefully harvested reason one after another to the speculative kernel of a founding premise or principle, then much of the time, especially if that time happens to be in short supply, the art of logic is lost on us. A carefully constructed argument that slowly gains shape and force is of much less appeal to our binary, black-and-white brains than an all-or-nothing haymaker: a knock-down affirmation that comes up and hits us right between the eyes.

We are drawn, involuntarily and inexorably, to prominent, fully formed, incontrovertibly heaped conclusions. With potentially disastrous consequences. The world of our ancestors may well have been black and white. But the colour of now is grey. We draw lines to create contrast because it is through the stark juxtaposition of contrast that we *see*. But the greater the contrast, the lesser the finer-grained detail. And the lesser the detail, the greater the potential for ignorance and errors of judgment.

CHAPTER 3

When Categories Collide

؛

*Light is meaningful only in relation to darkness, and truth
presupposes error. It is these mingled opposites which people
our life, which make it pungent, intoxicating. We only exist in
terms of this conflict, in the zone where black and white clash.*

LOUIS ARAGON

O N 13 NOVEMBER 1999, in Caesar's Palace, Las Vegas, the British boxer Lennox Lewis defeated the American Evander
Holyfield to become the undisputed heavyweight champion of the
world. The fight went the full twelve rounds with the judges awarding it to Lewis by unanimous decision. When the result was
announced the British contingent went nuts. The victory was particularly sweet because just eight months earlier, in Madison Square
Garden, New York, in a fight against the same opponent, their man
had been robbed of the title by a scandalously dodgy decision. This
time, however, there had been no messing about.

In a ring jam-packed with security, members of the press and
representatives from both camps, well-wishers thronged around
the dreadlocked, sweat-beaded Lewis to congratulate him. Many
of those present, including Holyfield's promoter, the self-styled
king of boxing Don King, were, not untraditionally, dressed in

black tie. Those who weren't, excluding trainers and corner officials, were soberly attired in business suits. But one man stood out like a sore thumb.

Frank Maloney, Lewis's diminutive, cheeky chappy manager, once described by Don King as a 'mental midget' and a 'pugilistic pygmy', was parading around in front of the capacity Colosseum crowd, and millions of other fight fans tuned in on Pay-per-View, in a rather different kind of suit. One decked out in the colours of the Union Jack.

The suit had become his trademark, one that he usually wore on boxing's big occasions. In many ways it fitted him perfectly, a sartorial summation of the man himself. Hard-nosed, thick-skinned, brash, fearless, shameless. Ever the showman. For over three decades, Frank really was one of the godfathers of the sport, equally at home in smoke-filled rooms, sweat-soaked gyms and the vampish clutches of scantily clad tabloid newspaper models, quick on the draw with the macho, locker-room banter and the show-stealing one-liners.

The suit was Frank all over.

A decade and a half on from that testosterone-fuelled night under the glare of the Las Vegas spotlights, a night when Frank became the first man ever to manage an undisputed British world heavyweight champion (a feat only recently equalled by Tyson Fury's manager, Mick Hennessy), I sit down with her for dinner at Chapter One restaurant in Kent. The Union Jack suit is now long gone. As are one or two other bits and bobs. Kellie, as she's known these days, is dressed in a red Donna Karan number with a sleek grey cardigan thrown over the top. The shoes, black with a modest heel, are Jimmy Choo, and the handbag, navy, Moschino. There isn't a hair out of place in her shoulder-length, strawberry blonde bob.

But the voice is still Frank: the wide boy, Cockney patter of eighties South London.

'I'd always felt that I was born in the wrong body,' she says. 'For as long as I could remember. But I was determined that I would never let it beat me. That was the way I thought of it. Like a fight. I was in the fight game and that's what I was up against – a dangerous opponent that I couldn't knock out but one that I could, hopefully, keep at arm's length.'

Of course, she couldn't tell anyone in the business.

'They'd have slaughtered me. They'd have had me slinking around the ring in a bikini at the beginning of each round, holding up a number. The only person I could talk to was a secret therapist I had on speed dial. I remember being really angry one day and shouting down the phone: "Don't you ever call me a transsexual!" and they said: "But you are, John!* You are!" and I said: "How do you know?" and they said: "Because you keep ringing me!"'

Kellie tells me that, as Frank, she used to go into women's shops, buy clothes, and then panic and throw them away. She wouldn't even try them on. The activity was a release of sorts. But the fear was being caught.

'Now that I'm a woman, most people have been all right about it. I remember the first time I ran into a rival promoter – we used to bang into each other quite a bit in the eighties and nineties – not long after I started gender reassignment. I was wearing a full-length dress, stilettoes and a wig. "Fuck me, Frank," he said. "You've changed!"'

I'd first met Kellie a year or so previously when we'd shared a stage together at a talkSPORT radio event in London. I surprised *her* with my analysis of Mayweather–Pacquiao and she surprised *me* with her background knowledge of psychopaths. ('That's thirty years in the boxing game for you!') After the event, she asked me what I was working on and I told her I was interested in how we categorize the world and how our brains put everything we see into little cognitive pigeonholes.

I mentioned Aristotle.

---

* Even on the phone Kellie didn't use her real name for fear of being found out.

Some two and a half thousand years ago the Greek philosopher had been among the first to write about categories and, in so doing, became the architect of what we now refer to as the classical theory of categorization. The theory consisted of four main premises but there were two that had particularly grabbed my attention.

First, in sharp contrast to the Sorites principle, Aristotle had proposed that categories are black and white; that there exist fixed and clearly defined boundaries between them, with no grey areas, no blurred lines. So any one individual star, to tweak our earlier astronomical analogy, in any given categorization sky, belongs unequivocally to one, and only one, of the proposed constellation categories within that sky. Or, in Sorites terms, there is no such thing as a 'between heap'. Just heaps and non-heaps.

Secondly, Aristotle had specified that categories were characterized by a set of 'necessary' properties that must be shared by all their members. In other words, in order to belong to a given category an item must tick every box on the feature checklist, must satisfy all of the entry criteria. And, moreover, that none of these criteria should be deemed more definitive of, or integral to, category membership than any of the others. All should be equally important.*

When I'd finished, Kellie had reached into her handbag and taken out a book. It was an advance copy of her autobiography

---

* The other two premises are that (1), all members of a category have equal membership status (i.e. a tomato is just as much a *fruit* as an apple); and (2), once all the necessary criteria for category inclusion have been established, then those criteria are 'sufficient' for category membership. No additional features are required. A simple example. Consider the possible 'necessary' and 'sufficient' features for inclusion in the category *singing*. Having a 'good voice' is clearly not necessary (Bob Dylan) but being able to stay in tune *is*. Having a good voice is not a sufficient criterion either because there are people out there who *do* have a good voice but are unable to stay in tune. Conclusion? The ability to stay in tune is both a necessary and sufficient feature for the category *singing* whereas having a good voice is neither.

bearing the distinctly *un*-Aristotelian title, *Frankly Kellie*. 'If you're after an alternative approach,' she'd said, slapping it on the table, 'you may find this useful.'

'You know, Kellie,' I'd said. 'You would've given Aristotle one hell of a headache.'

She'd giggled. 'I've given a lot of people headaches over the years, mate. No skin off my nose.'

She's not wrong. And now here we were, twelve months later, with dinner wearing on and our conversation entering the latter rounds of the evening, and poor old Aristotle and his classical theory of categories taking a bigger pummelling than Holyfield had in Vegas.

'Did you know,' Kellie inquires, 'that Facebook currently lists something like seventy different categories for gender?'

Yes, I say. I do.

'Well, then,' she says. 'Assuming that in the dim and distant past we started out with two – male and female – that's not exactly a ringing endorsement for the black-and-white, open-and-shut lobby, is it?'

No, I say. It isn't. But these days, I explain, the classical theory of categories has been largely debunked. And cognitive science, led, in the mid-1970s, primarily by the pioneering efforts of a dauntless, venturesome, jungle-bashing young psychologist at the University of Berkeley by the name of Eleanor Rosch, has long embraced the notion of categories being fuzzy around the edges. Even at the time Aristotle didn't have things all his own way. He may never have met him, or even have heard of him, but Eubulides' Sorites paradox would have given him a run for his money.

I provide the example of games, used by the Austrian-born philosopher Ludwig Wittgenstein in his 1953 book *Philosophical Investigations* to illustrate what he called the 'family resemblance' theory of categories. Wittgenstein invited his readers to call to mind the concept of games – board games, card games, ball games, children's games, *any* game – to see if they could discern something, anything, that they all had in common. His conclusion was that although various similarities *would* seem to exist it was difficult to nail them down. They appear and disappear in a complex maze of multidirectional meaning. Games, Wittgenstein proposed, do have common features, but there is no one feature that may be definitively found in all of them.

'I can think of no better expression to characterize these similarities,' he wrote, 'than "family resemblances", for the various resemblances between members of a family: build, features, colour of eyes, gait, temperament, etc. etc., overlap and crisscross in the same way. And I shall say: "games" form a family.'

But Kellie raises a good point. One, in fact, that Wittgenstein himself made at the time, and one, as we shall see shortly, that the intrepid Eleanor Rosch also picks up on.

'Just because categories aren't clear-cut doesn't mean to say that you don't know what's in them,' she observes. 'We might not be able to draw the exact line between games and whatever the opposite of games is . . . not-games? Or between "man" and "woman". But that's not the same thing as saying that games and men and women don't exist. We all know what a game is when we see one. Boxing's a great example. And we all know what a man or a woman is when we see one. Most of the time.'

In other words, our black-and-white brains might not be able to draw the line between a heap and a non-heap when we're adding one grain at a time. We might not be able to track and determine the granular, incremental, chimeric transition between them. But put a heap and a non-heap side by side in front of us and we instantly notice the difference. It's sufficiently black on the one hand. And sufficiently white on the other.

In 2016, Kellie tells me, as a case in point, the Boarding Schools Association in the UK issued guidelines to teachers on how to address students who identify as transgender or as a non-binary gender (those who see themselves as neither entirely male or entirely female).

The guidelines state that teachers should refer to transgender students as 'zie' rather than 'he' or 'she' and 'zie/zem' rather than 'him' or 'her'. 'Zie' was derived from an earlier pronoun, 'sie', which was ultimately rejected for sounding too feminine: 'sie' is German for 'she'.*

'But,' as Kellie mischievously remarks, 'we've still got "he" and "she". We still have men and women.'

I probe Kellie a bit further about the impact that womanning up has had on her life and on those around her.

'The kids will always call me dad,' she says. 'Because that's what I am. I'm their dad. They call me "dad in a frock". So to your way of thinking I may not be the best example of "dad" in the "dads" category. I'm a woman, for a start. And yes, I wear frocks. But I'm still in it. So that's anything *but* black and white.'

I mention a couple of cases that have been in the news: the Jewish transgender woman from Manchester who has been blocked from seeing her five children after a family court ruled that it would lead to both the children and her ex-wife being ostracized by their ultra-orthodox (Charedi) community and unable to live normal lives; the transgender woman who objects to being called her children's 'father' on their birth certificates because it violates her human rights – the word 'father', her lawyers argue, not only reveals her transgender status for all to see, it also serves as a permanent reminder of the man she never truly was.

Even the British Medical Association has entered the debate. In a recent memo to staff, employees are advised to refer to 'expectant mothers' as 'pregnant people' in order to avoid discriminating against intersex men and trans men who may also conceive.†

---

* In 1975, the Chicago Association of Business Communicators held a contest to find replacements for 'she', 'he', 'him', and 'her'. The competition was won by Christine M. Elverson of Illinois who suggested that the 'th' should be dropped from 'they/them/their' to create 'ey/em/eir'.

† The guide also suggests that someone who is 'biologically male or female' should be called 'assigned male or female' and that the phrases 'born man' or

'I think there are always going to be times when something's got to give,' Kellie says. 'And, by the same token, on some of those occasions greater public interest and having a consistent set of rules is going to be more important than individual rights. What about the rights of the kids to have a "father"? Will she ban them from calling her "dad" because it will bring back bad memories?'

Out on the street Kellie hails a cab. I hold the door open for her and she sweeps on to the back seat with the elegance and precision of one of Lennox's straight lefts.

The window comes down.

'So,' I say, 'three, five years' time . . . odds on a Union Jack dress?'

Kellie laughs. 'I've got great legs,' she says.

'born woman' should not be used in relation to trans people as such terminology is 'reductive and over-simplifies a complex subject'.

It continues: 'Gender inequality is reflected in traditional ideas about the roles of women and men. Though they have shifted over time, the assumptions and stereotypes that underpin those ideas are often deeply-rooted.'

News of the guide comes shortly after it emerged that a Briton who was born a girl but who has lived as a man for three years and is taking male hormones has put gender reassignment surgery on ice in order to have a baby. Dad-to-be Hayden Cross, twenty, who is legally male, will be the first man to give birth after asking the NHS to freeze his eggs before he completed his full transition in the hope that he might have children years later. He solicited the services of an anonymous sperm donor on the internet to become pregnant. Since then, Cross has had mixed emotions about his experience.

He told *The Sun* newspaper: 'I was happy but I also knew it would be backtracking on my transition. It's like I have given myself one thing, but taken away something else from myself in the meantime. It is a very female thing to carry a baby and it goes against everything I feel in my body.'

On the subject of the NHS refusing to pick up the tab for freezing his eggs, Cross added: 'It was like they were saying I shouldn't procreate because I am trans – it's not right . . . people think you can't be a man and have a baby but it's not that simple. This is my only chance. I want the baby to have the best. I'll be the greatest dad.'

Category wars

An hour or so west from Davis, California is Berkeley, home to another world-famous university. On the eastern shores of San Francisco Bay, it's where the I-80 from New York City finally ploughs into the Pacific. It's dusk as I pass the Richmond turn-off, the freeway lights of the city limits winking in the kaleidoscopic gloaming.

Back among the category labs of America's west coast I'm thinking about Lynn Kimsey and Lisa Oakes, of dogs, cats, butterflies and bugs. What had I learned from these taxonomic titans? Well, for starters it appeared that categorization was innate. That much seemed indisputable. And with good reason. If we lacked the ability to organize the world, to sort our experiences into cognitive clumps of shared semantic meaning, then everything around us would be chaos and nothing would be certain or predictable. We'd be forever caught up inside an eternal Sorites matrix. Then again, as I'd found out from Mike Anderson, there's a cap on the usefulness of such neurocognitive housekeeping. If those existential piles are arranged too neatly, are tidied up too fastidiously, our capacity to generalize from one context to another will be limited.

What I had learned, I eventually conclude, from Kimsey and Oakes and Anderson, is this. In order for categorization to be effective, to fulfil its evolutionary remit and simplify the world, there has to be a balance: maximum similarity between members of the same category on the one hand and maximum dissimilarity between members of different categories on the other.

Generalization, in other words, is good. But it can only go so far. The brain has to know when to stop.

The morning after I arrive in Berkeley I make my way over to the university. Waiting to meet me in a café called The Musical Offering is Professor Eleanor Rosch. In psychology, Rosch is a legend. In the mid-1970s, armed with little more than a multi-coloured paint chart, she disappeared into the jungles of Papua New Guinea to run an experiment* and by the time she came out had gathered a bunch of data that pretty much singlehandedly knocked Aristotle's classical theory of categorization clean out of the park.

Eubulides would most certainly have approved.

The Dugum Dani people of PNG's central lowlands have only two colour words in their lexicon. *Mola*, which translates as 'white-warm'. And *mili*, which translates as 'dark-cool'. They don't 'see' colour like we do. Yet when Rosch presented them with a carefully selected sub-set of her multi-coloured chips consisting of an eclectic, polychromatic mix of world-class colour exemplars (e.g. pillar box red) along with some shoddy, substandard alternatives (e.g. the red of the inside of a watermelon) and then later tested their memory for these hierarchically organized shades – by presenting them once again, only this time concealed amid a larger selection of chips – she made a startling and revelatory discovery. Despite the fact that the Dugum Dani lacked pretty much all of the so-called 'basic' colour terms common to most cultures – the reds, the greens, the yellows and the blues, for instance† – their memory for the genuine articles (the pillar box reds and Coke can reds) was better

---

* Rosch actually went equipped with a set of Munsell Colour chips. These are standardized chips, first designed by Albert H. Munsell, professor of art at what is now the Massachusetts College of Art and Design, in the early 1900s, which represent the three dimensions relevant to our perception of colour (hue, brightness and saturation). There are 329 chips in total: 320 chips represent 40 different hues, each divided into eight different levels of brightness. The remaining nine chips represent 'black', 'white', and seven levels of 'grey'. In designing the system, Munsell's aim was to systematically order the colour spectrum in three-dimensional space based not on subjective, lay classification but on the rigorous scientific examination of individuals' visual responses to colour. As such, he became the first colour investigator to separate hue, brightness and saturation into perceptually uniform and independent dimensions.

† In Chapter 7 we shall revisit the relationship between language and colour perception when we examine the principle of linguistic determinism, or, as it's sometimes known, the Sapir-Whorf hypothesis: briefly, that the structure of a

than their recall of the counterfeits and cheap imitations (the 'watermelon' corals and the russet red 'autumn leaf' rip-offs).

To Rosch, this signified something fundamental about the process of categorization which we'll come on to in a moment. But first, we need to understand the core principles of rudimentary colour vision. Put simply, we perceive colour as a continuum in three-dimensional space. One dimension is produced by hue, which is the actual colour itself. Another by brightness, which is how light or dark that colour is. And a third by saturation, which is how dull or vivid it is. The yellowness of a lemon, for instance, would be high in hue, high in brightness and high in saturation, whereas the yellowness of a fading bruise, in contrast, would be low on these three attributes.

There are no visible borders in the colour continuum yet we humans are able to differentiate between 7,500,000 colour shades. The reason that the rich, abbreviated palette of the basic or 'primary' colours jump out at us so readily – why, for example, we are able to distinguish the 'seven' colours of a rainbow* – all comes down to their innate colour talent, what they've got lurking in their colour DNA. The loud, shouty yellow of a lemon and the hot, brutal red of a pillar box or can of Coke represent major nodal intersections on the colour grid: electromagnetic metropolises where the dominant perceptual tribes of hue, brightness and saturation all come together in one big Newtonian melting pot. These are the super capitals of the colour world. The Londons. The New Yorks. The Beijings. The cities on the colour map that never sleep.

Which was precisely Rosch's point. That members of the Dugum Dani, despite the fact that they quite clearly conceptualize colour

---

language determines the way native speakers think, and, in some cases, quite literally 'see' the world.
* For an explanation of the difference between basic colours, primary colours and the colours of the spectrum, see Appendix I on p. 313.

in a very different way to the rest of us, exhibit a significantly better memory for these chromatic 'city centres' than for some of the more dreary, less vibrant, less happening colour suburbs – much as you or I might do if we were tested – suggested something profound, elementary and, in the field of cognitive science, utterly revolutionary at the time. In direct opposition to the creed of categorization laid down by Aristotle some two thousand years earlier – principally that categories are black and white, that they possess sharp, unequivocal boundaries and that members must all share the same equally definitive features – there are manifestly some colours that are better examples of a given category than others. There are Zone 1, inner city, concrete jungle blues. And dormitory, commuter, suburban blues. There are West End 'shamrock' and downtown 'emerald' greens. And then, as James Joyce once described the waters of Dublin Bay, there are the small-town 'snot' greens too. Moreover, this doesn't just apply to colours. The same can be said of anything. There are amateur colds, journeyman crises, wannabe trafficjams. And then we have the professionals. Flu. Coronavirus. Oxford on a Monday morning.

'Take red,' says Rosch as we sit in front of a laptop with hundreds of different coloured squares arranged on it in a spectral chequerboard sequence. 'There are genius reds, very good reds, OK reds, below average reds and poor reds. And there is usually a single red – which may differ from person to person – which tops one's personal rankings as the best possible example of red: the prototype. The further a red is from that prototype the less good it is as an example of that colour category.'

She pauses and runs her finger around some indeterminate hues in the shadowy spectroscopic hinterland between red and orange.

'Not only that,' she continues, 'but colour categories have fuzzy boundaries. Sometimes, around the edges, it's not clear exactly which colours are members of which category and which aren't.

Some are marginal. Take this square here, for instance. Is it red or is it orange?'

I move a bit closer to the screen. And then a bit further back. I'm really not sure. It's slap bang in the middle. It's not quite a heap of red. But then again, it's not a heap of orange either. It's a couple of wavelengths short.

'It's just like your boxing friend pointed out in relation to gender,' she continues. 'Just because you don't know for certain where the boundary between the sexes lies doesn't mean to say that men and women don't exist.'

The implications are considerable. Not least for the trans lobby.

I suggest to Rosch that 'man' and 'woman' therefore represent the prototypes, if you like, of the gender spectrum. Like 'red' and 'orange' and those other basic colours picked out by the Dugum Dani. And that when red becomes orange is when it starts to get kind of interesting.

'Well, that's one way of putting it,' she says, smiling. 'I think Wittgenstein once summed it up very nicely when he said that in the normal course of events two next-door neighbours don't need an exact boundary line to tell them on whose property they're standing. They just know.'

Rosch tells me about another study she conducted back in the seventies, this time involving furniture. She asked two hundred American college students to rate, on a scale of 1 to 7, whether they regarded a series of listed items as good examples of furniture. The items ranged from 'chair' and 'sofa', which both topped the rankings, to a bed (number 13), to a piano (number 35), all the way down to a telephone, bottom of the pile at number 60. In other words, just like 'colour space' there is also 'furniture space', plus an infinite variety of other spaces all with their inner city prototypes, leafy suburban satisfiers and backwater hinterlands. And this doesn't just pertain to objects, to material or physiological phenomena. It pertains, as Wittgenstein implied in his attempts to distil the essential properties of games through the notion of family resemblances, to ideas and concepts as well.

'A nice example I always use is a bird,' Rosch says. 'The classical model of categorization, Aristotle's model, was a model based on definition. So, for instance, a bird would be defined as something possessing certain features like feathers, a beak and the ability to fly, and for something to be categorized as a bird not only would all of those features have to be present but they would all have equal status. Which, of course, would have been a bit of a problem for an ostrich or a kiwi or the dodo.

'But in the prototype model the category *bird* would consist of different birds, some of which would be more birdlike, more prototypical, than others. A robin, for example, is more prototypical of a bird than, say, a penguin or a cassowary.'

Makes sense. But it's when we get into the conceptual realm that things start to get a bit more interesting. Doesn't a world that's becoming increasingly divorced from nailed-on, zero-sum solutions and ever more dependent upon least-worst options make an ideal hunting ground for prototype theory? Isn't the theory tailor-made for the intricacies, uncertainties and ambiguities of modern life: a halfway house between Sorites and Aristotle? Everything is on a continuum. But there are some shades of grey that are 'blacker' or 'whiter' than others?

Our brains evolved in a volatile, primitive environment in which binary categorizations of the world – fight or flight, approach or avoidance – were essential for survival. But several million years on, the survival game has changed and the categorization tables have turned. The sustained, progressive development of our higher-order faculties, which led, over time, to a densely packed curriculum of cultural and scientific advancement, has smudged the ones and zeros of our forebears' black-and-white world into an infinite horizon of grey-scale. Yet our chequerboard minds continue to divide us and rule us. Like fossilized light from an ancient, far-flung galaxy, our brains reflect the glimmerings of prehistory, with all the needs, demands and exigencies of our dark ancestral past.

Wittgenstein was right. But only up to a point. In the normal course of events categories are not a problem because, like next-door neighbours out in their proverbial backyards, we all 'know where we stand'. But life is more like tennis than gossiping over garden fences. Much of the pivotal action occurs not in the centre of the court but around the edges. On the lines. Where, more often than not, we *don't* know where we stand.

In tennis, of course, the only categories of any real significance are *in* or *out*, and games, sets, matches and championships can all hinge on the fractions of a millimetre that demarcate the boundary between the two. But in life the stakes are infinitely more complex and the margins immeasurably finer. Substitute *right* and *wrong* for *in* and *out*, for example, and you see the problem. On the precarious perimeters of the court of moral reasoning, a centimetre of conscience either way can easily make the difference between life and death itself. And with *right* or *wrong* there's no ontological Hawkeye to establish with absolute certainty, one way or the other, which it is. In fact, the closer the ball approximates to the line, the fuzzier the border between the two categories becomes, the more tentative, arbitrary and morally indeterminate the decision from the umpire's chair.

The less black and white the lines on the court appear.

The victim of a violent assault lies in a hospital bed in a deep coma. For three months they show not the slightest flicker of consciousness and remain on full life-support. At what point do they move from the category *alive* to the category *dead*? Moreover, at what point might their assailant be deemed culpable of their murder? In New Zealand at the time of writing, and in the UK until the mid-nineties, the law stipulates that if a victim dies from their injuries at any point up to a period of a year and a day after being subject to an attack, and it can be demonstrated that the cause of their death is directly related to that attack, then the perpetrator can be tried for their murder. This, for example, is why Chief Superintendent David Duckenfield, match commander on the day of the 1989 Hillsborough football stadium disaster, was charged with the manslaughter by gross negligence of only 95 people rather than 96 (the ninety-sixth victim having remained in a coma for almost four years until he died in hospital).

But what if, as with the Parkrun conundrum, we invoke Sorites? What might be the outcome if life-support is discontinued precisely 366 days and one second after it has been deemed necessary? Would the attacker escape a murder charge and instead face one of grievous bodily harm? And what might be the consequences of the plug being pulled a second or two earlier? In New Zealand, pre-1961, when the counsel of the gallows was still regularly sought over murder, and pre-1965 when the same was true in the UK, would their own lives have then hung in the balance?

At the other end of the ontogenetic spectrum we face a similar dilemma over the status of the unborn. At what point during development do we transition from the category *embryo* to that of *person*? Twenty-four weeks, when brain activity begins? Twenty-four weeks and one day? Twenty-three weeks? Twenty-eight weeks, as it stood half a century ago in 1967? Forty days for boys, ninety for girls as decreed by Aristotle? The moment of conception if you're a Catholic? Just as there exist no visual borders in the colour continuum, there exist no biological borders in the developmental continuum. An embryo matures gradually from single-celled zygote to newborn baby. There is no one defining foetal Big Bang moment when the singularity of 'personhood' irrevocably explodes within us.

Does it really make sense then, in the abortion debate, even to *attempt* to quantify the point on the prenatal spectrum at which 'personhood' might be deemed to begin? In the UK termination is legal up to twenty-four weeks. But what does that mean, exactly? Consider two women sitting next to each other on a train. Does it mean that the twenty-three-week, six-day-old foetus residing in the belly of one should somehow be regarded as less of a 'person' than that in the belly of the other, conceived, let's just say for the sake of argument, a mere twenty-four hours beforehand?

Are we not just bickering over single grains of sand?

Or consider the following scenario. A team of doctors battle to save the life of a premature baby born at twenty-three weeks while just a couple of doors down the corridor in the same hospital another woman is legally entitled to undergo an abortion on a foetus of the same gestation.

It's complex and unsettling. Yet live we must with such black-and-white distinctions. Such stark existential dichotomies. They are an inevitable consequence of the mismatch between us and reality; the terms and conditions of our inexorable, unending quest to navigate a world in which an imperceptibly gradated landscape must necessarily be perceived through a binary categorical lens.

Let's look at it another way. Does there exist, embedded deep within the template of our life span, some specially appointed hour, some solemnly significant day when a person who is categorized as 'middle-aged' suddenly gets categorized as 'old'? Or when someone who is categorized as 'young' suddenly becomes 'middle-aged'?

Clearly not.

Then why can't we accept that the opposite might also hold true: that at no specific point along the spectral in-utero timeline does a collection of cells, a rash of genetic stardust, suddenly erupt into a fully fledged human being?

The reason is because it matters. The life of the unborn is sometimes under threat. We are compelled to instantiate a hard, unconditional border between the categories *person* and *non-person* so that we can then introduce a further criminological airlock between the categories *murder* and *not-murder*. *Then* we can instigate yet another dichotomous distinction between the categories *right* and *wrong*. Why? Because we cannot make decisions of any shape or form without first arranging the cognitive raw material – the facts, the figures, the essential information – into categories.

The lives of the young, the middle-aged and the old, on the other hand, are equally protected under law. It is no less wrong to take the life of a fourteen-year-old than it is to take the life of a forty-four-year-old. Or, for that matter, even though they may be just as vulnerable as an infant or an unborn child, a 104-year-old. Once we are out of the womb age is of far less consequence. Relatively speaking, we're safe. There are no Eubulidean aspersions as to the metaphysical status of our being. No Sorites-style decisions based on our number of cells.

I tell Rosch about a tragic turn of events that occurred in Ireland several years ago. In October 2012, a thirty-one-year-old Indian dentist called Savita Halappanavar walked into a Galway hospital and requested a termination. She and her husband, Praveen, an engineer with the Boston Scientific company, had been resident in Ireland for four years at the time and she had suffered a miscarriage. Yet despite the fact that her uterus was badly ruptured, doctors at the hospital rejected her pleas for help. A scan revealed that the heart of the seventeen-week-old foetus was still beating. Abortion was illegal in the country.

The consequences were devastating. By the time the foetus' heartbeat eventually stopped, Halappanavar had contracted sepsis. Several days later she died.

Halappanavar's death sparked a storm of protest not just in Ireland but around the world. So much so that in May 2018 the Irish people voted overwhelmingly in a referendum to back new legislation allowing abortion in the predominantly Catholic nation to be performed under limited circumstances: should the mother's life be in danger were her pregnancy to continue, or were she to be suicidal.

'You are not aborting the child. You are only taking steps to save the mother of the child,' commented Fr Dominic Emmanuel, spokesman for the archdiocese of Delhi in the *Hindustan Times*, in a powerful example of 'framing', an influence technique that we'll be exploring a little bit later on in the book: how, often, it's not our actions per se that are important but rather how we see and interpret those actions.

But, of course, such legislation came too late for Savita Halappanavar and her husband. In any other country she would have been OK. But in Ireland, in 2012, a dose of Aristotle saw her pay the ultimate price. Rather than viewing the category *murder* as comprising a sliding scale of prototypical, good and less good examples of the act depending on context, the judiciary construed the category as consisting of any action in possession of the feature 'the taking of a human life', and the category *human life* as comprising any biological agency this side of conception.

Rosch shakes her head. But, she points out, the question of what we might call 'optimal generalization' – where one draws the line that demarks the greatest similarity between members of the same category and the greatest divergence between different categories – is always going to be trickier when it comes to concepts. Not least because concepts, as we shall see in the next chapter, incorporate attitudes, beliefs and moral and ideological convictions upon which the very foundations of one's social and psychological identity, one's holistic sense of self, are established.

Were one to draw a line between *red* and *orange* only for someone else to draw a slightly different line, what does it matter? One's need for validation, for clarity, consistency, certainty and meaning does not ordinarily extend to the categorization of colour hues. The composition of categories like *murder*, on the other hand, is a different proposition. Disagreement over category membership, over necessary features and the definitional geography of boundary posts, provokes anxiety, unease and, in some cases, overt hostility. It is an assault on the self. An incursion on who we are.

This gets me thinking. Seventy-something categories for gender. Can that really be right? Is *that* optimal? Does it really move us forward as a society? Does it deepen our grasp of reality, facilitate the wheels of social interaction and everyday decision-making as were the originally intended functions of our hardwired ability to categorize? Or, rather, does it do the opposite: throw a large taxonomic spanner in the works and complicate it unnecessarily?

Natural selection has programmed our brains in primal, ancestral binary. We code the world in terms of ones and zeros, and the more personal and significant the categorical landscape – gender, right now, is a good example – the greater our tendency to feel strongly about the matter. Yet when we're presented with a range of categorical information and must navigate the relevant options to assess, evaluate and ultimately commit to the numerous life choices that each of us face routinely day to day – What car should I buy? What holiday should I book? Which candidate should I vote for? – it is neither infinite variety nor bald, binary pot shots that we're looking for. Both confuse, confound and consistently lead to inferior, suboptimal decisions. What's required is a happy medium. No matter what kind of sand we're sifting, too many heaps is just as bad as too few.

To illustrate, consider the following taxonomy of the animal kingdom attributed by the twentieth-century Argentine essayist Jorge Luis Borges to an ancient Chinese encyclopaedia entitled the *Celestial Emporium of Benevolent Knowledge.**

> On those remote pages it is written that animals are divided into (a) those that belong to the Emperor; (b) embalmed ones; (c) those that are trained; (d) suckling pigs; (e) mermaids; (f) fabulous ones; (g) stray dogs; (h) those that are included in this classification; (i) those that tremble as if they were mad; (j) innumerable ones; (k)

* From 'The Analytical Language of John Wilkins', 1942.

those drawn with a very fine camel's hair brush; (l) others; (m) those that have just broken a flower vase; (n) those that resemble flies from a distance.

Remarkably, this eccentrically eclectic, proto-Linnaean system of classification never really cut the mustard. As a framework for zoological codification, its sublime organizational architecture never saw the light of day. The reason, of course, is that it's bonkers. Human categorization of the world is not the arbitrary product of historical accident or metaphysical whimsy. It's the evolutionary end point of a rigorous biological selection process, the aim of which is to derive maximum information from minimum cognitive effort. This is achieved, in stark contrast to the dotty demarcations of the Celestial Emporium, by reducing the infinite array of differences between all of the stuff 'out there' in the world around us to an optimal number that the brain can squeeze into the hand luggage of working memory as we strive to maintain a manageable, comfortable and portable portfolio of choices.

As well we should. Our brains may be big and hugely evolutionarily expensive but come with a severely restricted category allowance. There's a limit to the number of options that we can reliably hold in our minds. And it's actually surprisingly low. Too low, by far, as we shall discover in the following chapter, to make any kind of sense out of seventy different categories of gender. Or seventy different categories of *anything*.

Too many heaps in our everyday decision-making sandpits, too many squares on the black-and-white chequerboard of life, and our brains struggle to cope.

We're a lot more 'narrow minded' than we think.

CHAPTER 4

The Dark Side of Black and White

❛ ❜

If we had a keen vision and feeling of all ordinary human life it would be like hearing the grass grow and the squirrel's heartbeat, and we should die of that roar on the other side of the silence. As it is, the quickest of us walk about well wadded with stupidity.

GEORGE ELIOT

IN 1995, THE BRITISH actor Hugh Grant managed to piss off an entire Welsh community by pronouncing that it lived in the shadow of a hill and not a mountain. Ffynnon Garw, Grant claimed, fell slightly shy of the requisite height – 1,000 feet (305 metres) – necessary for the conferral of full mountain status and he had scientific evidence to prove it. He had scrambled to the top with a curmudgeonly artillery of sextants, theodolites and slide rules and had unleashed them on the summit. The numbers weren't good. Ffynnon Garw came up short. By sixteen feet. It wasn't a heap but a non-heap.

To say that Grant wasn't popular when he came back down would be putting it somewhat mildly. The villagers went crazy. And then they went *up* – to the top of the mountain (or should I say hill?) – to conduct a spot of sneaky topographical surgery. With

Grant still around, sleeping off his exertions in an upstairs room of the local village inn, Morgan the Goat, the owner of the hostelry, held court in the bar below. Dismissing the suggestion that they hasten Grant's demise on the grounds that someone, somewhere in Notting Hill would be bound to notice, he hatched a plan. If Ffynnon Garw wasn't a mountain then how about they *made* it one?

And so they did. With barrows, stones, sand, dirt and, needless to say, an ample supply of beer, they built an extension to the hill, added a few extra grains to the non-heap. And then they sent Grant up again.

He descended scratching his head.

Ffynnon Garw was a mountain after all.

The Englishman Who Went Up a Hill and Came Down a Mountain was, of course, a movie; a nineties British rom-com set against the backdrop of World War One; an offbeat, homespun study of the politics, psychology and backs-to-the-wall machinations of local community spirit. Grant played the blow-in cartographer, Reginald Anson, a man unwittingly intent on levelling not just the village mountain but also its self-esteem – fragile, debilitated and already significantly compromised by the tragedies of a vast, unfathomable combat unfolding far beyond the steeple-studded valleys in the fields and trenches of Europe.

But some twenty years after the film was first released, some very real Englishmen (and Welshmen) went up a very real Welsh hill, and . . . well, came down a very real hill. John Barnard, Graham Jackson and Myrddyn Phillips, who together comprise G&J Surveys, pronounced that Moelwyn Mawr North Ridge Top in the Snowdonia National Park no longer cut it as a mountain as it had failed to attain the *genuine* official qualifying height of 2,000 feet laid down by the altitude police.

Failed, that is, by a margin of just 0.9 inches. Or 2.3 centimetres.

Barnard, Jackson and Phillips are professional summit-busters who train the latest GPS technology on mountaintops in much the same way that traffic cops train speed guns on cars. Except they don't hide in bushes eating crisps. Their mission? To identify and expose shoddy mountainship and to thereby ensure that ramblers know precisely what they're getting for their windswept, rain-lashed footslogging. Their bible of mountain law is *The Mountains of England and Wales*, fastidiously compiled and painstakingly illustrated by John and Anne Nuttall.

Moelwyn Mawr's removal from this important tome had been on the cards for some time. *The Mountains of England and Wales* clearly stipulates that not only is there an entry-level altitude of 2,000 feet but there also needs to be a minimum differential of at least 15 metres (50 feet) between its highest point and any tract of land connecting it to its next most prominent neighbour.

The ridge top of Moelwyn Mawr stands at 2,132 feet. No problem there. But that is what we might call its 'absolute altitude', its height above sea level. When, however, the surveyors measured it against the adjacent hill, in this case the main peak of Moelwyn Mawr, it was a different story. Crucially, they found that the height difference between them fell short of the minimum 15 metres required to confirm its mountain identity. By, as I mentioned, just 2.3 centimetres.

'We haven't had one as close as this,' admitted Mr Barnard. 'The locals are not going to be pleased with us.'

He wasn't wrong. The community of Ffestiniog, which sits in the shadow of Moelwyn Mawr, is a historic mining town with a population of around 5,000. It relies heavily on tourists, who come to visit the heritage railway, the Llechwedd Slate Caverns, and the *mountains*. As things stand, *The Mountains of England and Wales* currently lists 189 mountains in Wales that are higher than 2,000 feet and which are also in possession of the officially accredited increment of a 15-metre peak. Having one to call your own is a serious business. The peaks have even got a special name, 'Nuttalls', after the book's authors.

'People will be disappointed to hear that what they always knew as a mountain has been downgraded to a hill,' commented local guesthouse owner Richard Hope, a sentiment shared by many of the town's residents. 'There must be a certain degree of error in these things – 23 millimetres could be a mound of dirt. All it would take is for someone to do what they did in the film and grab a shovel and pile up a bit of rock.'

And make a mountain out of a moel hill?

Managing the catalogue of life

The human capacity to categorize is as old as the hills. In fact, without it we wouldn't, strictly speaking, have 'hills' at all because that, as we now know, constitutes a categorization in itself. Similarly, we wouldn't have 'old' either. Because that constitutes another. Without the ability to categorize, all the books in the labyrinthine library of life would be housed, in no particular order, on random, ramshackle shelves with not the slightest rhyme or reason as to their arrangement.

For every tome we needed to lay our hands on we'd be compelled, in a sea of seamless sameness, to start our quest from scratch. We might have the *words* for carrot, or cucumber, or courgette. But without the ability to categorize such items as 'food' we wouldn't have the faintest idea what to do with them. We wouldn't know whether to eat them, smoke them or write with them. Everything would be a matter of trial and error, derived from First Principles. Prediction would be flat-out impossible. Without the ability to describe, distinguish and discriminate, the world as we know it would simply cease to exist. Everything, simultaneously, would signify both something and nothing. Everything, at once, would be equally important and unimportant. Nothing, because everything, would be salient.

To illustrate, let's stick with the library analogy and consider the plight of a frazzled, overworked librarian whose job it is to scurry up and down ladders all day long scouring the shelves, volume by volume, row by row, in an incessant quest for whatever manuscripts her patrons might solicit. It wouldn't be long before the library closed down. Sure, she *could* get lucky and the sought-after book might well be among the very first ones that she happened to pick up. On a good day, if she got really lucky, it might even be *the* first. But chances are it wouldn't be. In fact, Sod's Law would dictate that it'd be among the last.

Unless, that is, she manages to acquire a system that enables her to refine her marathon search odysseys so that, no sooner is she handed the title of the book, she can immediately take the stairs to the correct floor of the library, head straight to the correct area of that floor, walk right up to the correct bookcase, quickly peruse the correct shelf in that bookcase, and then select it from a handful of other books on the same, or similar topics, as opposed to a bibliographic haystack of apocalyptic proportions. Imagine how much easier her life would be then!

But *now* let's imagine how such a system might evolve over time. Arguably the first line our overstretched librarian might draw in the bibliological sand would be that between fact and fiction. She might house factual tomes in one part of the library. Works of fiction in another. Once that categorical distinction has been made she might then set about further subdividing each of these two fundamental 'supergenres' into a series of subsidiary categories. Fiction, for example, she might carve up into – amongst other sections – romance, crime and horror. Fact, on the other hand, she might partition into the arts, humanities and science.

Supplementary compartmentalization would then proceed along increasingly more nuanced lines. In science she might establish four related but autonomous provinces: the *natural sciences* (the investigation of natural phenomena with the aim of discovering the laws, precepts and principles behind them); the *formal sciences* (the study, such as that of mathematics, computer science and information theory, of the rules and properties inherent in formal systems like logic, linguistics and calculus); the *social sciences* (the study of human behaviour and societies, such as anthropology, social psychology and economics); and the *applied sciences*, such as medicine and engineering, which focus on the practical applications of scientific knowledge.

Our librarian might then further partition the natural sciences into the *physical sciences* and the *life sciences*,* and subsequently decide to *really* fine-tune her taxonomy. She might, for example, take earth science and divide its shelf space into eight constituent sections: atmospheric sciences, ecology, geology, geophysics, glaciology, hydrology, physical geography and soil science. She might then take each of these eight academic sub-disciplines and apportion them into a series of even more rarefied research fields. Geology, for instance, she might separate into mineralogy, petrology and paleontology. Atmospheric sciences into climatology and meteorology, etc. Until eventually, after as many distillations as it takes, experience tells her that she has finally arrived at 'peak classification'; that she has alighted upon the optimal number of categories to enable her to retrieve, accurately and efficiently, pretty much any scientific title the outside world can throw at her.

Lynn Kimsey, with her jungle of drawers full of immaculately pinned creepy-crawlies, would be proud.

The guiding principle underlying the need to categorize is the same for all of us. It is a fundamental need to tame the feral uncertainty of the world; to domesticate grey into house-trained black and white. We are compelled, through the necessity of simplification, to

---

* Broadly speaking, the physical sciences investigate the behavioural properties of non-living systems and include such topics as physics, space science and earth science, while the life sciences – biology, zoology and botany, for example – specifically involve the biologically based study of living organisms.

collate the ambiguity of what's 'out there' and to assimilate its infinite randomness within the finite, systematized shelf space of what's 'in here': the codified cubbyholes of ever-increasing nuance deep inside our brains. But concealed within this process of forensic categorization lurks a fundamental dilemma. How do we know when to stop? How do we know when enough is enough – when we've reached the point at the other end of the orderly–disorderly spectrum when our shelves are optimally classified and when further categorization becomes impractical and counterproductive?

For real librarians the question poses little difficulty. Sophisticated analysis of bibliographic citation indices, together with systematic cross-referencing procedures, permit them to identify, with the aid of cold, hard, number-crunching algorithms, the precise point in the specialization spiral at which the well of ever increasing nuance runs dry.

But what about the rest of us? The division of gradated realities into optimally stacked shelves of categorical distinctiveness foreshadows profound, far-reaching consequences in more areas of our lives than we might otherwise care to realize.

Just ask the people of Ffestiniog.

A famous study conducted several years ago in America demonstrates this beautifully. Customers in a gourmet food store were presented with two displays of jam and were handed a coupon entitling them to a $1 discount if they made a purchase. One display contained twenty-four jams. The other just six. 30 per cent of the customers who eyed up the smaller selection bought a pot compared to only 3 per cent of those presented with the larger range, though the larger range attracted the greater interest.

The reason?

Something that proponents of decision science refer to as the 'tyranny of choice': the greater the number of alternatives the more difficult we find it to make up our minds. Because the more we keep comparing like with like.

The psychological fallout is inevitable. And by no means beneficial. Faced with an array of too many categories to choose from we
begin to feel incompetent. It's a drain on the memory – 'What was
the Tequila, Lime and Strawberry Conserve like again? I know it
was good but I can't remember why!' – and we end up making worse
decisions than if the field was more restricted.

Which is obvious when you think about it. But wholly insignificant when you don't. The only way to break the cycle is to go for
the Nike option. Just do it! To bully ourselves into making a choice
that we might otherwise not have made had the range of possibilities been narrower and the spread of alternatives less disparate.
Then, of course, we beat ourselves up for our trouble.

The drawing of lines is essential. The drawing of lines is inevitable. But at some point or other we need to draw a line.

Tainting by numbers

Seven is a magic number. You hear that often enough. But is there
really any truth in it? If you're into the science of categories it seems
that there actually might be. Consider, for a moment, the following.

There is an infinite array of colours out there. And yet how many
of these does the electromagnetic piñata of composite white light
reveal upon deconstruction? Just seven. 'Richard Of York Gave
Battle In Vain' as many of us will remember from school: red,
orange, yellow, green, blue, indigo and violet.

There has been an infinite array of musical notes woofed,
oinked, mooed and miaowed in nature's production suite over the
years. And yet (in Western music at least) how many basic notes –
sharps, flats and octaves aside – are available for use? Just seven: A,
B, C, D, E, F and G.

The human face is capable of producing hundreds of different
micro-expressions. And yet ethnological research from all four corners of the world has shown that only seven basic emotions are

recognized universally across all cultures: happiness, sadness, anger, disgust, fear, surprise and contempt.*

The Babylonians gave us the seven-day week, appropriated, by the ancient Judaists in Genesis, to frame the story of the Creation.

And then we have popular culture. The Seven Wonders of the World, Snow White and the Seven Dwarves, The Magnificent Seven . . .

Each of these categories, selected, in turn, from our sensory, cognitive, emotional, social and everyday working environments, comprises seven constituent members. But why? What did the Babylonians know all those years ago that we still subscribe to today?

Back in the 1950s, the Harvard psychologist George Miller provided the answer to this question in what, it turned out, was to prove the first empirical demonstration of a phenomenon now universally accepted by pretty much everyone in cognitive science as solid psychological gold. In a series of experiments, Miller demonstrated that there is a limit to the number of items that the brain can store in its short-term memory database at any one time.

And that number, you've guessed it, is seven. Plus or minus two.

We may, as we shall see in just a moment, be able to train ourselves to remember more items through the use of simple but effective memory-enhancing techniques such as grouping and association. But when it comes to the unschooled mind, no matter who we are or what we do, no matter how creative, or logical, or intelligent we might be, the maximum number of items from any one category that we can hold in our short-term memories at any particular time is fixed between five and nine. With seven being the average for most of us.

---

* It is acknowledged that there are some in the field of emotion science who would take issue with this view. For an alternative perspective see Lisa Feldman Barrett's excellent book *How Emotions Are Made*.

To demonstrate, try holding the following numerical sequence in mind:

6 5 1 9 5 4 3 5 9 4

If we interpret this as an arbitrary string of ten separate digits then it exceeds the permitted download capacity of short-term memory and the task becomes difficult. Ten categories (or 'chunks', as memory experts call them) are simply too many for our brains to keep track of all at once. But if we're able to discern two meaningful super-categories among the ostensibly random series of numbers (on 6 May, 1954, Sir Roger Bannister ran the world's first sub-four-minute mile: 3:59.4) then we reduce the information string to only two categories or chunks. As soon as we snap our fingers the original ten digits come running.*

Which begs an obvious question. Does this 'magic' number seven, as Miller himself described it in the title of his seminal paper, represent the optimal number of categories for *everything*, no matter what the composition of that range or the nature of those alternatives?

The answer, pretty much, is yes – at least insofar as our judgment and decision-making goes. In psychometrics, for example, between five and nine – seven being right in the middle – represents, according to most studies, the optimal number of grading points on rating scales: 1. Agree very strongly; 2. Agree strongly; 3. Agree moderately; 4. Neither agree nor disagree; 5. Disagree moderately; 6. Disagree strongly; 7. Disagree very strongly. Seven, it is thought, offers the perfect balance between the sensitivity to discriminate subtle, yet important differences in the variable being measured on the one hand and response simplicity on the other.

---

* Ever wondered why telephone numbers around the world generally contain between 6 and 11 digits, and that those digits are then typically subdivided into groups of between 2 and 6 digits? Now you know.

In the analyses of the fundamental building blocks and core characteristics that define us as human beings and the societies in which we live, the essential component criteria rarely exceed seven dimensions. The structure of personality, for instance, can be reduced to just five key variables – openness to experience, conscientiousness, extroversion, agreeableness and neuroticism – while the sociological dimensions upon which cultures might vary can be reduced to just six: power distance, uncertainty avoidance, individualism versus collectivism, masculinity versus femininity, long term versus short term orientation, and indulgence versus restraint.*

And then we have straightforward judgment tasks like the one we just saw with the jams. Underlying the 'tyranny of choice' that bedevils complex comparisons are constraints on our working memory. Research shows, for example, that if we're required to compare and then grade order a set of stimuli – imagine, for the sake of argument, that I were to play you a set of tones on a keyboard and then, once I had played them, ask you to rank them in order of pitch – you'd do well until you got to around seven but then your performance would deteriorate dramatically.

Why? Because any more than seven and it's too much for your memory to carry around unaided. Items start spilling out of its clutches on to the floor in the same way that items of shopping might topple out of *your* arms on to the floor on a trip around the supermarket if you pick up too many.

* '*Power Distance* relate[s] to the different solutions to the basic problem of human inequality; *Uncertainty Avoidance* relate[s] to the level of stress in a society in the face of an unknown future; *Individualism* versus *Collectivism* relate[s] to the integration of individuals into primary groups; *Masculinity* versus *Femininity* relate[s] to the division of emotional roles between women and men; *Long Term* versus *Short Term Orientation* relate[s] to the choice of focus for people's efforts: the future or the present and past; and *Indulgence* versus *Restraint* relate[s] to the gratification versus control of basic human desires related to enjoying life' (from Hofstede, 2011, p. 8).

Of course, if we place those items of shopping in a basket we can carry significantly more of them – in exactly the same way, as we've just seen, that if we place cognitive or perceptual stimuli in 'memory baskets', or chunks, we can remember significantly more. But some stimuli sit better than others in those baskets. Numbers fit fine. But sounds, smells, tastes – stimuli, in other words, that are more sensual than cognitive in nature – pose a little bit more of a challenge.

Which means that if we're confronted with an instrument consisting of more than seven items and we don't have a basket to put them in – twenty-four pots of jam, for instance – then the process of assessment becomes trickier. Our brains pig out on the buffet of different options and we struggle to digest the slurry of available comparisons. Cognitive load – mental effort – increases and the decision-making gas gets turned up higher and higher. We stew. We simmer. We bubble. Until eventually, to prevent our synapses from boiling over completely, we turn off the gas and go simply with what we know. Hunches. Heuristics. Suspicions. Whatever pot we might have in our hands at the time.

Black and white is good. But not if it comes in more than seven shades.

The link between the anatomy of rating scales and our preferences for various food items doesn't exactly jump out at us. Who would've thought that something as arbitrary and innocuous as the number of categorical response items on a survey could have such profound implications for the jams and the jellies, the preserves and the marmalades that we see on our supermarket shelves? The manner in which we categorize can have far-reaching and deep-seated consequences for the way we live our lives. More so, as we've discovered, than at first we might expect.

Sometimes however, unlike with the jam example, we know exactly what we're doing when we categorize. And sometimes our motives are distinctly less than honourable. In the ambiguous world of social interaction categories can be armed, weaponized and deployed against others to perform any number of disreputable and self-serving deeds. And I'm not just talking about the big stuff: the cultural and ideological war crimes like apartheid and religious sectarianism that we see playing out on the news. I'm talking about the small stuff, the petty stuff, the behind closed doors, skullduggerous office-politics-type stuff that goes on in schools, town councils, hospitals, businesses – you name it – all day, every day, every*where*. Obscurity, irrelevance, ignominy . . . there's no limit to the number of credit-busting pigeonholes of doom into which we can categorize our enemies and competitors should either the need or inclination arise, as arise they do with remarkably convenient and casually unerring frequency.

Tailored job descriptions are a good example. It's by no means unusual for organizations to harbour their own preferred candidates for vacancies within their ranks and to put out ads in which both the essential and desired criteria for a post are fiendishly crafted to the qualities and CVs of the particular individuals in question. Granted, sometimes this can end in a maelstrom of NDAs. But at other times it can be highly amusing.

As an example of the latter, I once remember seeing a notice displayed on a college information board in Oxford alerting students to the fact that there was a modest bursary up for grabs aimed at subsidizing the costs of second-year geological field trips. The ad, printed on college headed notepaper and signed by the master, went on to stipulate that successful applicants should be 'from Liverpool or surrounding areas, into Metallica and Call of Duty', and have either the names 'Ben' or 'Murray', ideally both, clearly delineated on their birth certificates.

Now that's class.

It was an elegant way of notifying the head-banging, console-twiddling Scouser, Ben Murray – who would almost certainly have been the *only* second-year geologist resident in the college at the time, and who was probably a little hard up – that the fellowship were watching his back.

But it's not always like that. Less magnanimous were the actions of a certain head of department I once knew who took the rather unusual decision of downgrading from 'centre' to 'group' any departmental research unit whose complement consisted of less than four professors – thus rendering the label 'centre' for larger scale bodies and collaboratives. His reasoning at the time appeared sound enough. Standardizing in-house nomenclature would help streamline the outward appearance of the department. And, in so doing, would better manage the expectations of interested parties whose first port of call was its website. Pedantic, perhaps, but most people could see where he was coming from. Besides, the fallout from the change was minimal. Only one unit, in fact, was affected. So it was by no means the end of the world.

Then a year or so later an announcement went out to the staff. The department was to take ownership of a new building. Estates teams were hovering and floor plans were under review. As with all such manoeuvres there was the usual competition for space. Some lab suites and offices amounted to prime, much sought-after real estate. Others were just glorified stationery cupboards. One actually *was* a stationery cupboard. All of which precipitated that time-honoured logistical nightmare: how to decide on a fair and equitable basis, in the absence of prejudice or partisanship, who got what?

It was, indeed, a difficult and delicate conundrum. But one, it transpired, that the head proved more than equal to. After reflecting at length on the complex and sensitive matter he regretfully arrived at a blindingly obvious conclusion. Perhaps it made sense that first through the doors of the gleaming, state-of-the-art new building should be the leaders of the fledgling 'centres' while those heading up the 'groups' took their places in the queue behind. It seemed like the perfect solution. Logical, reasonable and devoid of any bias.

So that, pretty much, was that. The wise and kindly shepherd divided up his flock and all the lambs were happy. All, that is, except one, the black sheep of the department, who, a year or so earlier, had seen his distinguished if diminutive research unit demoted from 'centre' to 'group'. Which, it emerged, was also around the time that he and the head of department had had a bit of a falling out. And around the same time, too, that the head had first seen the floor plans.

Coincidence? Perhaps. But I've never really bought it. More likely that the head knew what he was doing all along; that it was score-settling jiggery-pokery from the start.

Categories can be booby-trapped. And categories can arrive bearing gifts. But much of the time they're neither good nor ill, they're just everyday boxes in which we deposit our lives. Which isn't to say that we should just leave them lying around. Boxes have issues, too. Too many and they get in the way. They trip us up, make it difficult to move and manoeuvre. Too few and we lump life together.

We can't taste the jams for the jars.

The Goldilocks principle

We're all hoarders to some extent. The 'you never know' mindset is hardwired. You may not have a significant proportion of the 1980s teetering around your living room or rammed halfway up the stairs. But what's that little pile of papers doing over there by the computer on your desk? Or that growing stack of magazines on the kitchen table? Chances are they're both outstaying their welcome because you can't decide what to do with them. Should you throw them away? Or should you hang on to them? You never know, just in case . . .

When most of us think of hoarders we think of Fagin-fingered, swivel-eyed hermits waist deep in washed-up washing-up-liquid bottles and Charles and Diana commemorative memorabilia. The virtuosos. The professionals. The elite of the hoarding establishment. So if we're all on a spectrum, what is it, exactly, that separates them from us? What's the difference between hoarding with a small 'h' and hoarding with a capital 'H'?

The answer may come as a surprise. Far from being crazy, or lazy, or both, psychologists believe that hoarding behaviour on a grand scale may be partially explained by a categorization disorder. More specifically, a so-called 'underinclusive' style of categorization in which individuals define categories more narrowly, therefore producing a relatively larger number but with fewer component members.

It's easy to see how such a mindset might work. Individuals with this kind of deficit may well find their possessions more difficult to group into categories when organizing their homes, perceiving fewer commonalities between them. In turn, they may, as a result, perceive items to be more unique, less replaceable, and ultimately more difficult to discard. In a worst-case scenario where every household object forms its own independent category no system of organization at all can be implemented. Just like our frazzled librarian before she performed her sorting surgery, everything is equally salient. And nothing can be thrown away. Hoarders see the world in *shards* of black and white. Not shades.

Research into people suffering from obsessive-compulsive disorder would seem to support this theory. One study, for example, pitted individuals with OCD against healthy volunteers on something called an 'Essentials' task in which participants are presented with a concept (e.g., a book) and are asked to select words that are considered essential for defining that concept (e.g., cover, pictures, pages, print).

Results showed that those individuals with OCD were much stricter than healthy individuals when defining category membership, selecting significantly fewer words and exhibiting a preference for those terms relating to prototypical category attributes as opposed to more peripheral attributes. In the case of 'book', for example, they opted for 'pages' rather than 'print' or 'cover'. *Under-inclusive* thinkers, in other words, are harder to please. In their eyes, for something to qualify as a book it would need to have more than just a cover, print or pictures. It would need to have actual pages.

But if over-structuring the world – being too heavy on the categorization gas – can cause us problems then the opposite, putting the brakes on too hard and too soon, can do the same. And the consequences can be even worse.

Compared to their under-inclusive counterparts, *over-inclusive* thinkers may well demonstrate unduly relaxed criteria when defining category boundaries and consider *all* things with pictures, pages, print and covers to be books. But they may also do the same with *people* and consider all Muslims as extremists, all Westerners as infidels, all people who support the minimum wage as Communists, all individuals who argue in support of keeping the Cecil Rhodes statue at Oxford University as white supremacists, or all black males under the age of thirty-five as gun-toting, knife-wielding crackheads.

All of a sudden, it's a very different story. And we're into stereotyping airspace.

I'm reminded of an incident that took place several years ago on a train I was on in London. It was the last train out and a guard appeared in the carriage checking tickets. The gentleman sitting in front of me, a young, Harry Potter-reading vicar, had, for what I'm sure would have been a perfectly legitimate reason, evidently yet to purchase one. He waits, cash in hand, for the guard to make his way over.

'Now,' he announces breezily, with the inspector looming above him, 'I've got a little story to tell you.'

The inspector shakes his head. 'Don't bother, mate,' he says. 'I already know the ending.'

The padre looks perplexed. 'Really?' he stammers. 'What's that?'

'A £200 fine,' says the guard.

It's funny, I know, looking back on it. But if hoarding is under-inclusion then stereotyping is over-inclusion, the other side of the categorization coin. Stereotypes are categories that don't know when to stop. To the guard those passengers not in possession of a valid ticket were 'all the same'. They were fare-dodgers. Freeloaders. Criminals. It was way beyond the realms of his psychological remit to countenance even the remotest possibility that among the wantonly ticketless there may well have been one or two hapless individuals with genuine reasons to account for their predicament.* If hoarders think in shards of black and white then extremists think in pure, unadulterated binary.

Us and them.

We are born to categorize and pigeonhole. And, just like everything else in life, we can either under- or overdo it. Get it right and we can pull off miracles. We can convict killers from bugs that are trapped and squashed in radiators. We can lay our fingers on books in libraries housing millions of volumes.

But get it wrong and we're capable of epically imprudent misjudgments. And like our theodolite-wielding mountain-whisperer Hugh Grant, we're brought back down to earth with a seismically heavy and eye-opening cognitive bump.

---

* One of my editors, whose opinion I greatly value, makes an interesting point about this anecdote which, I must admit, I hadn't previously considered. A story that is supposed to be about the perils of stereotyping seems to have fallen into its own trap, she suggests. The gentleman in question could well have been a criminal dressed as a vicar. Or an actual vicar who was dishonest. In a way, she continues, the guard wasn't stereotyping at all. His job remit doesn't allow for him to make exceptions. He must treat everyone without a ticket the same, and while someone's appearance – a vicar – could lead to stereotyping (i.e. 'trustworthy') he actually chose *not* to do this. This is a brilliant counter-analysis of the situation and, of course, could well be true. Having been there, however, there was definitely something about the guard's demeanour which alluded to a less generous interpretation of events . . . or, perhaps, simply to my own stereotype of jobsworth ticket inspectors.

CHAPTER 5

The Viewfinder Principle

*The earth is round and flat at the same time. This is obvious.
That it is round appears indisputable; that it is flat is our
common experience, also indisputable. The globe does not
supersede the map; the map does not distort the globe.*

JEANETTE WINTERSON

SOMETIMES IN LIFE it's good to stand alone. Sometimes it's good to let other people in. Sometimes it's good to focus on the finer details, up close, sequentially, one at a time. Sometimes it's good to take a few steps back and get a sense of the bigger picture.

It's where we set our viewfinder that's the key.

Many years ago, back in the 1920s, an elderly farmer was eking out a living on the Russia–Finland border. The farmer was Finnish but the frontier between the two countries cut right through the middle of his farm. Not just his farm, in fact, but his living room. If he sat by the dresser he was in Russia. If he sat by the fire he was in Finland.

One day two government officials, one from Finland, the other from Russia, drop in on him unexpectedly.

'On which side of the border do you wish to live?' they ask him. 'There is a census coming up and we need to know who is eligible.'

The farmer weighs up his options. It's a devilishly difficult dilemma. He is Finnish by birth but the Russian authorities have been good to him over the years, providing him with free running water and regular repairs to his outbuildings.

Eventually, after much deliberation, and not wanting to give offence to the Russian official, the farmer comes to a decision.

'I have been very grateful for the support that Mother Russia has given me over the years,' he explains. 'She has helped me through the toughest of times. However, on reflection, and with a heavy heart, I think it best that I live out my days in Finland.

'You see, I'm an old man now and can't take the cold like I used to.

'Another Russian winter might kill me!'

Black-and-white thinking is complex and multifaceted. In this understated masterclass in emergency, black-belt diplomacy ten feet to the left or right is, quite literally, neither here nor there to the Russian and Finnish authorities. But to the elderly Finnish farmer it is everything. Such margins are the fulcrums upon which the highs and lows of everyday life get their name. Slim, small, and seemingly inconsequential, they surround us at every turn. Let's consider a couple of examples.

In the UK, driving or attempting to drive whilst above the limit of 0.08% BAC (blood alcohol concentration) is illegal.* If you blow into the tube and your result is 0.08% then you're OK. The light stays green, you drive back home, and pour yourself another. If, on the other hand, your blood comes out at 0.09%, then you are not OK. The light turns red. You get into the police van and you have a cup of coffee down at the station. Just that measly 0.01% can make all the difference, not just to how your night turns out but

* The figure is 0.05% in Scotland.

perhaps to how your life turns out. You might, for example, not have been heading home. You might have been driving, instead, to the bedside of your wife or partner who was about to give birth to your first child. Or, alternatively, was about to breathe their last. Such petty, minuscule margins. Such pervasive, momentous consequences.

But can it really be any other way? Imagine two people driving along a dual carriageway one night when they're pulled over by the police and given random breath tests. Both turn out to be fractionally over the limit with readings of 0.09% BAC. One of them hands over his keys and takes a seat in the back of the squad car.

The other is given a friendly slap on the wrist.

'It's a bit of a grey area, and we're only talking 0.01%,' says the officer. 'So just keep an eye on it, OK?'

Same readings. Same levels of intoxication. Different cops drawing very different lines.

Let's revisit the Coronavirus crisis and the UK government bailout for the self-employed. On 26 March 2020, the chancellor, Rishi Sunak, announced that the government would pay self-employed people who had been adversely affected by COVID-19 a taxable grant – worth 80 per cent of their average monthly profits over the last three years – of up to £2,500 a month. Applicable to 95 per cent of self-employed workers the plan, nevertheless, was capped at a threshold of £50,000. Which meant that if my average profit over the last three years came out at £50,000 and yours at £49,999 then you would benefit while I'd end up with zilch. Fair? Maybe, maybe not. But again, can it really be any other way?

Or take sport. In soccer, the recent introduction of the Video Assistant Referee (VAR) has seen goals disallowed because the players who scored them were as little as a single invasive millimetre offside. But, some will argue, and have done quite vociferously, does one millimetre *really* count as offside? 'It's ruining the game' claims one TV pundit. 'If rules are rules then anything that helps to enforce them can only be a good thing' contends another.

Big picture versus finer details. Or, as data scientists refer to them, actionable insights. Do we zoom in, pan out, or hover in the middle distance? Billboards of black and white? Or pixels of black and white? It's the age-old problem.

Sport, in fact, is one of the arenas in which the positioning of our personal viewfinders is most fascinating and, needless to say, where its repercussions are most keenly felt. Not just when it comes to performing. But, crucially, when it comes to preparing. Not long after I returned from America and my west coast odyssey around the category labs of northern California I got talking to Ronnie O'Sullivan, five-time world snooker champion and, in most people's eyes, the greatest player to ever pick up a cue, about the importance of fanatical, absolutist, all-or-nothing tunnel vision when it comes to achieving at the very highest levels of elite performance. Then, a couple of days after Ronnie, in the ExCel Centre in London, I did a corporate gig for the World Travel Market with Seb Coe, the double Olympic 1500 metres champion of 1980 and 1984, multiple world-record holder, and now the current president of the International Association of Athletics Federations (IAAF), and had a similar conversation with him.

Turned out I was in for a surprise. Somewhat against expectation, both Seb and Ronnie were actually rather circumspect about the benefits of black-and-white thinking when preparing for big events, the pixelated variety especially. Yes, it can have its advantages. But it can also have serious side effects. Too much focus and you start to lose perspective. And when you start to lose perspective . . . you also start to lose.

'I'm at my best when the snooker table is over there in the corner of my life,' Ronnie tells me, as we gaze out of a first-floor room on to a dark, wet, traffic-light-smeared street in the middle of Halifax. 'When it's at the centre there's just no avoiding it. You're always banging into it. Walking round it. Trying to get past it. That just does my head in.

'I'm not saying it's not important. Because it is. But snooker's funny. Sometimes it gets airs and graces – big ideas – and thinks it's more important than it is. This title. That title. This record. That record. Which is why tucking it away in the corner is good. I can go over to it whenever I want, for as long as I want. And it doesn't get in the way of everything else.

'I always play better when I *want* to play better. And to want to play *better* you've first got to want to *play*. And to want to play I can't be eating and drinking it. Because if you eat and drink anything that much – relationships, fitness, diet – all the other stuff you want to do goes right out the window. It becomes like an addiction. And eventually you just get sick of it.'

I'm talking to Ronnie in the green room of the Victoria Theatre, directly above the stage. Downstairs, later on, he'll be doing a turn in front of five hundred people. Diehard fans who've travelled from all over the north of England – Leeds, Manchester, Bradford and York – to spend a couple of hours in his company.

Over in the corner is John Virgo, known to millions of snooker aficionados across the world who watch the sport on TV as the voice of the game. Ironic, laconic and holding a beaker of vodka and tonic, he stands puffing on a Marlboro Light beneath a 60-watt energy-saving light bulb. 'But you've got to put the hours in though, Ronnie,' he offers, arm out of the window.

Ronnie agrees. 'Oh, yeah,' he says. 'Don't get me wrong. When I'm on fire and knocking 'em in I go into a kind of tunnel-vision mode. It's just me and the table and I block everything else out. And when you're preparing for a tournament you definitely need to be focused. Get your head down over the practice table. Have a routine and all that. But I guess what I'm saying is, the routine can't become an obsession. Not for me, anyway. It has to leave room for other stuff. Otherwise I'm like: what's the point?'

Seb Coe, as I mentioned, is of a similar view. On the banks of the Royal Victoria Dock in London's East End, we sip hard, thin coffee from corrugated cardboard cups. We're due on stage in half an hour's time and have stepped outside for a breath of fresh air. We forget we're in Canning Town.

'The only time I truly shut everything out and didn't allow myself to think about anything else at all apart from my performance on the track was in the build-up to the 1980 Olympic Games in Moscow,' he reflects. 'And we all know how *that* turned out. As history attests, it definitely didn't do me any favours. In fact, I think one of the reasons I was able to turn the tables so quickly after what happened in the 800 metres – and go on to win gold in the 1500 metres – was because I had people like Brendan [Foster] and Daley [Thompson] around to help draw back the curtains on my dark night of the soul and get me out of myself.*

'Four years later, in Los Angeles, I wasn't so myopic in my approach and it paid off. As much, I have to say, by dint of circumstance as anything else. In the early part of 1983 I'd contracted a blood disorder, toxoplasmosis, which pretty much affected my performance for the whole year. By the end of it I was hardly able to even get out of an armchair let alone run 800 metres. As late as March the following year I was still jogging around the track at Haringey Athletic Club with fourteen-year-old kids. So you might say *that* gave a modicum of perspective! It was definitely a bit of a leveller whichever way you look at it.

'But it was also a conscious decision on my part. If I found myself

---

* Brendan Foster and Daley Thompson are both legends of British athletics. Brendan, the founder of the Great North Run, currently the biggest half-marathon in the world, took Olympic bronze in the 10,000 metres in the Montreal games of 1976, having previously been crowned European champion over 5,000 metres in Rome in 1974. At the following Olympic games, in Moscow in 1980, Daley became Olympic decathlon champion and went on to successfully defend his title in Los Angeles in 1984. He set the world record for the event four times.

in a new city somewhere and wasn't training I'd go out and visit an art gallery, or a jazz club or a record shop, rather than adopt that siege mentality and board myself up in my room. And, as I say, it paid dividends. Not only did I enjoy the Games a lot more, I was simply beaten [in the 800m] by a better man on the day. I'm not saying that Steve [Ovett] didn't deserve to win in Moscow. But let's just say that in the race itself I didn't exactly do everything in my power to impede his chances!

'In fact, the first words my old man – who was also my coach – said to me no sooner had I stepped off the track were, "You ran like a . . ." Well, never mind what he said. But he was right!'

For historians of track and field, and for pretty much anyone in the UK over the age of fifty, the summer of 1980, and in particular the events that unfolded over those last few days of July and that first day of August in the Central Lenin Stadium in Moscow, will be forever etched in the memory. Sebastian Coe and his then arch rival Steve Ovett, the two best exponents of middle-distance running in the world at the time, and now widely regarded as two of the greatest of *all* time, were going head to head in the finals of both the 800 metres and 1500 metres.

For the Great British public and sports aficionados the world over it was a truly mouth-watering prospect, rendered more enrapturing still by the hypnotic thrall of the binary. You were either a Coe person or an Ovett person. There was no middle ground. Coe with his linear majesty and incorruptible form. Ovett, the barrel-chested bruiser, rough around the edges. It was Federer–Nadal. Ali–Frazier. The Beatles and the Stones.

It finished honours even. Steve took the 800 metres. Seb got his own back in the 1500. Each stole the other's thunder.

Seb tells a story that illustrates just how important it is not to get too cocooned up in yourself as an elite sportsperson, that espouses the benefits of companionable, joined-up living. The morning after he'd lost the 800 metres his friend and newly crowned Olympic decathlon champion, Daley Thompson, barrels into his bedroom unannounced. He bounds over to the window, yanks up the blinds and stands there looking out.

Seb's in a daze. 'What's the weather like?' is the best he can do under the circumstances.

Daley turns around with the cheekiest of grins. 'Oh, you know,' he announces nonchalantly. 'It all looks a bit silver to me!'

'That,' Seb says, 'was exactly what I needed. It was a call to arms. The beginning of the fight back. They say every cloud has a silver lining. But I wanted one with a gold lining!'

And, he adds, it was precisely because he had friends like Daley and Brendan Foster there next to him, shoulder to shoulder in the trenches, that he *did* fight back; that he was able to shove an industrial-sized spanner in the works of the 1500 metres several days later and take gold to his nemesis's bronze. Daley and Brendan became not so much the light at the *end* of the tunnel as the light *inside* it. They subtly changed the dynamic. From black and white to black interspersed with white.

Ronnie has a similar thing going with the artist Damien Hirst and Rolling Stones bassist Ronnie Wood. Wood's face is often alighted upon in the crowd by the TV cameras at major tournaments. And Damien, when he's not spectating, will usually hang out in the dressing room, larking around, shooting the breeze and generally lending an air of anarchic, eccentric normality to the highly charged proceedings.

'Damien loves his snooker,' Ronnie tells me, as, back in Halifax, we shimmy around the cold, sunless right-angles of the steel and concrete stairwell down to the theatre. He and John are about to go on and it's time to break a leg. 'But when I'm playing in a tournament and he's there with me,' he continues, 'it's never about the snooker. It's about me being happy. About me having the right head on. About me being me, really.'

But Seb has another story, one which provides a telling insight into the madness that lurks on the other side of the coin, within the obsessional confines of those blind, intemperate *dark* squares on the black-and-white-thinking chequerboard. It's a story that shows just what his mentality was like back then, what it *had* to be like back then, in the build-up to the Moscow Games.

It was Christmas morning, 1979 – raw, bleak, cryogenically cold – and he'd just finished quarrying twelve sleety miles out of an implacably brutal endurance colliery known as the Peak District, a rugged, rocky and, on this particular morning, refrigerated National Park in the south of the north of England. But, lounging in front of the TV after dinner, the turkey and mince pies beginning to pick up the pace towards a mid-afternoon siesta, he starts to feel uneasy. At first, he isn't sure why. Then, all of a sudden, it dawns on him.

'I looked out of the window, at the leaden sky and the empty street and the howling gale whistling through the trees, and thought to myself: "You know what? I bet Steve's getting blown about out there somewhere putting in his second shift of the day."'

So that was that. He pulled on his tracksuit, headed back out and clocked up another seven miles in the wind and the snow and the ice.

Many years later, in 2006, the two old adversaries ran into each other again at the Commonwealth Games in Melbourne. With the past now firmly behind them and the rivalry that had once electrified a nation long since switched off at the mains, they sat down together for dinner. Seb recounted the tale. Steve coughed.

'You mean, you only went out twice on Christmas Day?' he said.

In art as in life

Personal viewfinders have a variety of different settings. On the one hand, there's the wider social context of 'big picture versus little picture mindset' that we've been considering with Seb and Ronnie.

Then, on the other, we have the more granular, more cognitive perspective of under- and over-categorizing that we encountered in the previous chapter.

Zoom in too close to what we're trying to look at – as Seb did in 1980 and as Ronnie sometimes does – and we can't see the wood for the trees. We can't, to return to our chess analogy of earlier, see the board for the black and white squares. That, as we discovered, is the perplexingly paralysing problem that hoarders have to deal with. Everything is unique, standalone, special – newspapers, washing-up-liquid bottles, duvet covers, bath mats and toasters – so nothing, not even the chattels of selfhood and personal identity, can be thrown away. That's right – when Facebook endorses seventy-plus categories of gender, that's hoarding!

Zoom out too far, in contrast, and we can't see the trees for the wood. We lose the nuances, the subtleties, the finer-grained details of the object or scene in front of us and have trouble distinguishing it from similar objects or scenes.

That's when we get into stereotyping.

The message couldn't be simpler. Peering through the viewfinder on an inappropriate setting is not a good idea. Not only is the picture out of focus, it generates considerable stress and anxiety.

And it's not as if the viewfinder needs to be 'out' by a lot to cause us problems. Even a seemingly insignificant miscalibration can set the cat amongst the pigeonholes and create unseen and unwelcome difficulties. Consider, for example, at a trade fair, the difference between presenting customers with six pots of jam and twenty-four pots of jam in a taste test. It might seem trivial if you're on the marketing side of things. But not, as we've discovered, if you're one of the manufacturers. Just a simple tweak here and there of the flexible psychometric viewfinder could make the crucial difference between getting that contract or not. Between your product being one of the favourites, or anonymity through overcomparison.

To borrow a rudimentary observation from the art world, it's all about achieving a true and accurate balance between the nature of the picture we're looking at and the distance we maintain from the canvas. It's about perceiving the brushstrokes at the optimal level of compositional detail to enable us to envision, as best we possibly can, the painting as a whole. The principles of optimal categorization may well apply to physical objects and perceptual stimuli like colours and furniture, as we learned from Eleanor Rosch in Chapter 3. But they can also apply in the social environment, too. To how we manage, structure and organize our lives.

Take, as an example, two iconic works: *Impression, Sunrise* by Claude Monet, painted in 1872, which can be found in Musée Marmottan Monet, Paris – or, of course, online – and *The Girl with a Wine Glass* by Johannes Vermeer, c. 1659–1660, which you can see in the Herzog Anton Ulrich-Museum, Brunswick. Imagine that these two paintings are being exhibited right next to each other in a gallery and that you are standing in front of them at a point equidistant between the pair, surveying each of the scenes in turn. Do you think that would be the best way of doing it? Do you think that by viewing them that way, rooted firmly to the spot in the one intermediate position, you'd be best placed to appreciate the extraordinary degree of industry, creativity and craftsmanship inherent in each of the canvases? Of course not. You'd need to move forwards and backwards. Shuffle from side to side. Twiddle with the settings on your art appreciation viewfinder . . . up, down, down, up . . . to determine the optimal perceptual range.

Impressionist works such as Monet's are referred to as such for a reason. Proponents of the Impressionist school of painting generally create 'impressions' of reality using broader, brasher brushstrokes and dabs, slabs and squalls of colour that 'add up' to reality, that make most sense from a distance. In contrast, the paintings of an artist like Vermeer are better appreciated in comparatively greater proximity, where the deftness of touch and inestimable attention to detail may be subjected to appropriate scrutiny.

In his painting, Vermeer achieved the effect of radicalizing both light and texture on the lining of the young suitor's cloak through the use of a technique called *pointillé*, the application of paint to canvas not through continuous, contiguous movements of the brush but rather through the necklacing of loose, luminous dots in gleaming, celestial layers that appears to ensnare the very weave and wobble of the light. But if you want to see it you've got to get up close. If you're halfway across the gallery then forget it.

Sometimes it's good to zoom right in. To see the details, the specifics, the minutiae. Sometimes it's good to pan right out. And take in the bigger picture. Not just in art but in everything. In business, politics, society. In pretty much any sphere of life you care to mention.

Eddie Howe, the manager of AFC Bournemouth, is a hugely popular figure on the south coast of England – well, certain parts of it anyway – who has steered the football club from the brink of administration (he took up the mantle back in 2009 with the club languishing at the bottom of League 2 on a tally of -17 points for failing to comply with the Football League's insolvency rules) to Premier League success. As it stands at the moment he is currently the Premier League's longest-serving manager. Hopefully, at the time of going to press, he still will be.

I was invited to the Vitality Stadium, where Bournemouth play, by their chairman, the dapper and irrepressible Jeff Mostyn, to deliver the keynote speech at the annual Directors' Dinner. Jeff had his own part to play in Bournemouth's fairy-tale success story. Back in 2008, with the club literally five minutes away from liquidation, he wrote out a cheque for £100,000 to snuff out an unpaid tax bill. It marked the beginning of what followed with Eddie. What the Cherries call their 'Great Escape'.

Sitting in the boardroom the morning after my speech I spot Eddie and Jeff in the corner having a chat. The Bournemouth fans still have a chant: 'Eddie had a dream on minus seventeen . . .' and if ever there was a pair of big-picture fellas I was looking at them. What would they make of the viewfinder principle: the idea that although we're programmed to think in black and white sometimes it's good to think in slabs and chunks and sometimes in houndstooth check?

'Makes perfect sense,' says Jeff. 'And if you're the chairman or manager of a Premier League football club these days you need to have a very flexible viewfinder, I can tell you!

'During the course of a season you're going to have a lot of ups and downs. And although, ultimately, it's about the big picture – how many points you have at the end of it – you take every day, every week, every game as it comes. I know it's a cliché, but it's true. Sometimes you don't need the big picture or a goal in life. You just need to know what you're going to do next. In football it's the same. August to May is a string of *nows*. You just join up the dots—'

'—and hope that the shape you've got at the end of it doesn't look like relegation!' interjects Eddie.

We laugh. But Jeff's right, Eddie continues. 'What is it they say? A journey of a thousand miles begins with a single footstep. Well, that's true. But both the goal and the process have to be right. Yes, the devil *can* be in the details. But he can also be in the big picture. If you get the vision wrong you can be right every step of the way on your thousand-mile journey until the last one when you turn around and suddenly find that you're not where you wanted to be.

'You need to keep your head down. But you also need to look up every now and again. With the lads here at Bournemouth we strike a healthy balance in that regard. We divide the season up into a series of four-game groups, or "mini seasons", and assess where we are, how many points we've got, at the end of each one. You're kind of anchored in the present but also focused on the future at the same time.

'Plus, as Jeff said, it can be a long old season at times and dividing it up into shorter periods helps keep the lads motivated to either build on what they've done or, if results haven't gone so well, to regroup and go again.'

The message from the south coast of England is loud and clear. The artist and poet William Blake might have been able to see the world in a grain of sand and eternity in an hour. But he never had to stave off the prospect of relegation to the Championship every season. Or run the gauntlet of the annual shareholders' meeting. For most of us it's one or the other. Either a matrix of pixelated *nows*. Or a widescreen, wall-mounted *later*. And no matter what game you play you need both if you want to keep playing it.

At the opposite end of the country to Eddie Howe and Jeff Mostyn in Bournemouth there is a psychological giant of a man whose voice natural selection somehow managed to chisel out of the leftovers of the Rocky Mountains. Sean Dyche runs the show at Burnley Football Club and, after Eddie, is the Premier League's longest-serving manager. There isn't much Sean doesn't know about putting together teams and setting them on a win cycle, and, like Eddie and Jeff, he also sees merit in the viewfinder principle. A couple of seasons ago I did a talk up in Burnley for Dyche and the team and afterwards we went out to an Indian restaurant for dinner.

'We have a saying up here,' Sean tells me, as the first of the night's major signings, the garlic chilli prawns, steps out on to the table. 'The minimum requirement is maximum commitment. It's not, at the end of the day, about winning or losing. Well, it is . . . but you know what I mean. It's about working your bollocks off every single day. Giving it everything. And I mean everything. And then, when you've got nothing left, giving what's left.

'Legs. Hearts. Minds. You've clocked it up on the wall at Barnfield [Burnley's training centre]. And it's what I drum into players all the time. To put in a shift over ninety minutes all three have to be working together.

'The way I see it is like this: if results are the small print then the headlines are the team. They're the big picture. And, to be fair, the fans and the club are the even bigger picture beyond that. And as a manager you've got to keep an eye on all of it.

'And you know what one of the main barriers to seeing the big picture is? Ego. That's why nights like this, curry and a few beers, are so important. It makes sure everyone's feet, including mine, stay exactly where they should be. On the ground. And you keep it simple. To win a football game, you've got it all right there. No need to overthink or overcomplicate it. Legs. Hearts. Minds.'

Burnley Football Club has some interesting fans. And Sean thought it might be good for me to meet one of them in particular. What Sean knows about win cycles, Alastair Campbell – the former Labour Party stalwart – knows about spin cycles. Not the ones you select on your washing machine. The ones programmed into the machinery of politics and power. Alastair's CV reads like an episode list from *House of Cards*. Director of Communications for Tony Blair's Labour Government during the 'cool Britannia' years of the late nineties and early noughties, and before that both the Downing Street press secretary and Blair's spokesman and 1997 election campaign director, he now serves as the chief interrogator for *GQ* magazine, editor at large for *The New European* and ambassador for a stack of mental health campaigns. What Alastair doesn't know about politics, media and campaigns just isn't worth leaking.

Sean's instincts are spot on and I'm glad of the introduction. If ever there were a place where appropriate levels of black-and-white thinking take centre stage, where viewfinder settings are important, where they need, quite literally, to be kept an eye on, it's politics. The ball started rolling, or the division bell tolling, in France, in the eighteenth century. On 5 May 1789, the French Revolution commenced with a meeting of the Estates General, a legislative and consultative assembly representing the French estates of the realm. Summoned to Versailles by King Louis XVI, members of three social classes – the First Estate, the clergy; the Second Estate, the nobility; and the Third Estate, the commoners – all came together in an attempt to contain France's escalating financial crisis.* For reasons that still elude historians to this day, during the course of the convocation those estates that sided with the King – the nobility and higher clergy – took up positions to his right, while those who opposed him – the commoners and middle classes – perched on his left. From that moment on, the 'left–right' metaphor for political orientation was born, inaugurating an enduring, dualist vocabulary of 'right-wing' Conservatives, 'left-leaning' Democrats and 'centrist', 'middle of the road' Liberals.

Precisely why this metaphor has stood the test of time and has extended its polemical reach beyond the end-stage infrastructure of the *ancien régime* to democratic systems of governance the world over should by now, of course, come as little surprise. It is for exactly the same reason that the metaphor 'heaven' and 'hell' has survived for so long as a trope for good and evil and, as we've gradually been discovering, why the terms 'male' and 'female' have, until recently, had it all their own way in pretty much every tongue and dialect in the history of human expression. Political persuasion, like morality, like gender, like race, is a messy, complicated business. Were we, for instance – as we did with skin colour in the Introduction – to stand all possible colours of the political spectrum in a line, from the reddest of the red at one end to the bluest of the blue at the other, through wine, mauve, magenta, plum and purple, and were to make our way along it, we'd be faced with a similar continuum of indeterminate intermediates. Red and blue – left and right – are the prototypes of the political world. They are

* In the late 1700s, France was in the midst of a crippling financial crisis as a result of the enormous debt that had ensued from French involvement in both the Seven Years War (1756–63) and the American Revolution (1775–83). Corruption, together with the indulgent lifestyle of the royal family and the French court at Versailles, did little to mitigate this calamitous state of affairs.

nodes of ideological intersection where the ties that bind are perceived to be at their strongest and conviction abides at its purest.

Red and blue are the black and white of politics.*

As I ramble on, Alastair listens politely. I considered it prudent
to sketch a bit of background as to the broader sociological provenance of my thesis but suddenly realize halfway through that
recounting the origins of party political division to one of the greatest political strategists of modern times is like trying to explain fire
to Prometheus. Alastair is better informed than most on the art
of black-and-white thinking. On how the way we categorize – the
volume, the amplitude and the frequency of black-and-white
dichotomies – holds sway over how we decide. The viewfinder
principle just couldn't be more up his alley; the enduring psychological tension between party political binaries on the one hand
and individual differences in narrative, conviction and personal
ideology on the other has been the object of his attention for years.

'Politicians are human beings the same as everyone else,' he
tells me, over fish and chips and tap water. 'The big difference is
the scrutiny and the importance of the decisions for so many other
people. And because of the historic left–right divisions and the role
of parties we force the impression that Party A all think one thing
and Party B all think another. But it is more complicated than that.
People on different sides often agree more than they pretend.'

His words, given their context, are exquisitely well timed.
Spoken over lunch in the garden of a gastropub in Highgate, North
London, they come just a day or two after his much publicized
expulsion from the Labour Party. His crime? Voting for the Liberal
Democrats in the European elections in protest at Labour's then

---

* Alastair has spoken and written at length about his mental breakdown in the
1980s and revealed that mid psychosis he became fixated on, and paranoid about,
the words 'left' and 'right' and the colours red and blue. These concepts, he maintains, clearly drill deep into political minds.

stance on a second referendum over Brexit. He hasn't lost any sleep over it. 'I'm still Labour,' he smiles, and is confident, he tells me, that the party will eventually discover something 'more modern and sensible than Corbynism'.*

Alastair explains that politics sometimes feels like the M25, the invariably overcrowded and pungently flatulent motorway that orbits London. Just as you've got your three lanes, left, right and centre, on a motorway, so, he articulates, you've got your three main political lanes facilitating the flow of traffic in the democratic process. The analogy is a good one. It speaks to the viewfinder principle in a way that I hadn't expected. Lane discipline works when traffic is moving freely, when the road is clear and there is natural variation in speed. But when things go awry it suddenly becomes less important. When a snarl-up occurs everything descends into lockdown and road markings go out of the window. Unfettered by lane, drivers tend to do what they want. They chop and change. They weave in and out at will.

'Or take football,' he suggests. 'Whenever I go to watch Burnley I want them to win. Obviously. But that's not to say that I can't appreciate quality from opposing sides when it happens. Or a nice bit of skill by an opposing player. Or feel sorry when one of them goes off after a bad tackle. Or that if I were in charge of picking the England team all the English boys from Burnley would be in it and Kane, Sterling and Alexander-Arnold would be sitting it out on the bench. That's not football. And it's not politics, either. Not good politics, anyway. In fact, it's everything that's bad about it – and there's a lot of that kind of thing doing the rounds at the

---

* 'Corbynism' is a broad and nebulous term used to describe a cluster ideology of leftist political thought that coalesced and gathered steam under the Labour leadership of Jeremy Corbyn 2015–2019. Core principles include anti-imperialism, foreign non-interventionism, hardcore anti-capitalism, social liberalism, and participatory youth 'movementism'.

moment in the Trump–Johnson era. Cronyism. Intolerance of anyone other than full-blown sycophants. A deliberate undermining of truth.'

He's right. Viewed from a distance – whether as City fans, United fans, politicians, traffic wardens or Extinction Rebellion activists – people, like paintings, appear finished articles. On the basis of just fleeting impressions we either like or don't like, we either accept or reject, what we see. And we navigate the rooms in the gallery of life accordingly. It's only when we get closer up, focus on the countless different brushstrokes – the direction, the texture, the colour composition of the unique personality in front of us – that we begin to understand the work that's gone into each piece. Of course, we still may not like them. They still may not be our cup of tea. But at least we can appreciate the positives. The craft, the vision, the individual qualities of the artist.

Back in the summer of 2017, this is precisely what Laura Pidcock, the then newly appointed Labour MP for North West Durham, a Labour Party stronghold, *didn't* do in an interview with the left-wing online news site *The Skwawkbox*. The interview, at the time, made quite a splash in the UK press. The 'opposition' is a term widely used in politics to refer to members of the party or parties that are not the ruling party. In other words, that are not in power. As such, it is functionally innocuous. But during her interview Ms Pidcock pushed the envelope further. On being asked about her own personal take on the status of non-parliamentary, out-of-hours, cross-party relations between her fellow House of Commons MPs, she commented that she could never be friends with a Tory because, as she put it, they're 'the enemy'.

In Ms Pidcock's eyes, it emerged, there are two types of Tories. Those 'so blinded by their own privilege' that they are totally incapable of empathizing with anyone less fortunate than themselves. And those 'completely ideologically driven' and myopically in thrall to the barbarous panacea of capitalism.

'Whatever type they are, I have absolutely no intention of being friends with any of them,' she revealed. 'The idea that they're *not* the enemy is simply delusional when you see the effect they have on people – a nation where lots of people live in a constant state of fear whether they even have enough to eat.'

There were friends, she explained, that she *chose* to spend time with and none of them frequented the other side of the House. For her, she declared, it was 'visceral'. Being 'cosy' held little interest.

Ms Pidcock's comments garnered quite a bit of attention in the media. Much of it, to be fair, falling short of a ringing endorsement. But it was the response from those of her own political persuasion that was most interesting. Even *they* were quick to dust down the diplomatic bargepoles. The consensus was clear. The polarity of her terse, truculent, truncated taxonomy of Tories was, in most people's eyes, ridiculous. Although the fomentation of an either/or, all-or-nothing divide, championed by enmity and bolstered by ideological incomprehension may well carry with it a feeling of tribalist satisfaction – and, of course, be good for votes – it isn't, in the overall scheme of things, actually all that practical. It isn't, as Alastair Campbell observes, conducive to what politics, at the end of the day, is all about. Getting things done.

For that, one needs to liaise and build alliances. Sometimes with those who don't just *appear* to have the wrong end of the stick, but with those who really *do*. And, moreover, not just with those who have the wrong end. But also with those who have the wrong bloody stick entirely.

What Ms Pidcock perhaps failed to appreciate when she sat down for that interview with *Skwawkbox* was that her viewfinder was stuck on 'over-inclusive' landscape. She was thinking in black and white. And it was big, bad, brutalist black and white. Had she possessed, at the time, the desire or capacity to release it on to close-up; to twiddle the dials so that she could zoom in and out at will; to draw not just the one overarching, all-defining line between herself and her fellow MPs but rather a number of smaller lines demarking various subsidiary categories of interpersonal inclusion, then she may well, quite literally, have seen things rather differently. The 'enemy', most probably, would have been easier on her eye. Less skanky. Less samey. Less altogether alien.*

We've all seen those pictures in newspapers and magazines. Wood lice and dust mites and other infinitesimal monstrosities blown up a million times. The head of a flea, under the sub-atomic gaze of an electron microscope, is a thing of wonder. Same as the abdomen of a house spider and the eye of a cockroach. Even cancer cells appear beautiful at these extreme, subterranean levels of magnification.

So what am I saying? That we should jump for joy every time the specialist clears his throat? Or when we're sat in the doctor's surgery and it's 'not good news, I'm afraid'? Of course not. Everything in life is relative. Compassion included. At times like these we need our viewfinders in as far as they will go. On the trade stall of personal extinction the black-and-white taste test should consist of just two simple pots. One labelled 'life'. The other 'death'. Stereotyping all cancer cells as bad . . . is good. Just as it was good for our prehistoric ancestors to stereotype all rustles in the bushes as potentially lethal. These are things that could legitimately do us in. The gloves are off. The game is on. Empathy and discernment should quickly be shown the door.

But if, as Alastair points out, we're trying to sort out Brexit, or endeavouring to eradicate anti-Islamic or anti-Semitic sentiment from cultures and communities fragmented by fear and ignorance, then stereotyping whole herds of people as the 'enemy' – Labour, Conservative, the Far Right, Muslims, Jews, or anyone else for that matter – is a touch misguided and not entirely helpful. Black and

---

* In the 2019 parliamentary election Laura Pidcock lost her seat to Conservative Richard Holden.

white needs to be fractal, not binomial. Our viewfinders need to be *out*. Right out.

Until we can't see the 'them' for the 'us'.

On the subject of relativity here's a simple, practical and ever so slightly spooky demonstration of the viewfinder principle in action on which to end. Take a look at the picture of Albert Einstein shown below in Figure 5.1.

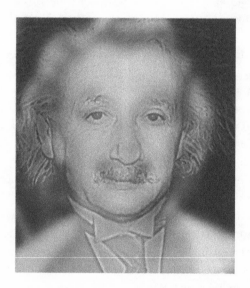

Figure 5.1: Marilyn Einstein.

But . . . *is* it Einstein?

That very much depends on how far away from it you are. Hold the shot at arm's length as you're probably doing right now and you'll be wondering what I'm talking about. But prop the book open on a table and view the same shot from across the other side of the room and you should, if you're like most people, suddenly 'get the picture'.

It's a powerful illustration of what most psychologists and hard-line psychophysicists would consider *the* founding principle of all human perception, from the fundamental mechanics of basic visual processes right the way up to the time-honoured ways in which families, faiths, football fans, political parties, even nation states, evaluate and interact with each other: *what* we see depends very much on *how* we see.

The illusion works because there exists a sizeable imbalance in the level of pixelation between the two respective images that make it up. The image of Marilyn Monroe comprises substantially fewer pixels than that of Albert Einstein and so can only be deciphered from a distance. At closer range, in other words, Marilyn is perceptually overpowered by Albert who, pixel-wise, is beefier and thicker-set, enabling minute features of his appearance – the strands of his hair and the wisps of his moustache, for instance – to jump out at the observer.

Speed, too, also plays a part. The researchers at the Massachusetts Institute of Technology who created the illusion presented it to people in their lab for a variety of different intervals, asking them what they saw. A clear pattern emerged. Those who were exposed to the image for a shorter length of time (30 milliseconds) saw Marilyn and nothing else. In contrast, those who were shown the image for 150 milliseconds were able to unscramble Albert.

The significance, from an evolutionary point of view, is profound. When time is short we see in black and white. Our brains, quite literally, check out the bigger picture at the expense of the perceptual small print. Which they had to, of course, in the days of our primitive ancestors, otherwise Albert and Marilyn would never have been with us to start with. There would've been no blondes. Nor gentlemen who preferred them. And no relativity either.

But cognitively, socially and culturally, too, the implications of the illusion are themselves comprehensively pixelated. Up close our brains discern the finer details. Like Einstein's wrinkles and famous walrus moustache. But as the distance increases so, in contrast, does our ability to nuance *decrease*. Until whoever we're looking at changes beyond all recognition. Across the other side of the table, Albert Einstein becomes Marilyn Monroe. Across the other side of the House, the Tories become the untouchables. And across the other side of the street, the Chens and the Chans, the Khans and the Kumars and the Kowalskis become the unreachables.

CHAPTER 6

The Complexity of Simplicity

: :

Take time to deliberate. But when the time for action has arrived, stop thinking and go in.

NAPOLEON BONAPARTE

W HEN I STUDIED applied maths at school – rather briefly, I lasted about an hour – the teacher began by setting us a well-known problem. Well known *now*, that is, thanks to Google. But not so well known at the time. The conundrum, commonly referred to as the 'Weight Problem of Bachet de Meziriac' after the French mathematician Claude-Gaspard Bachet de Meziriac who first conceived of it in his eclectic and exuberant manuscript of 1624, *Problèmes, plaisants et délectables qui se font par les nombres*, goes something like this.

A spice merchant's assistant is driving along the road in his truck one day when he hits the brakes to avoid an oncoming vehicle. The truck screeches to a halt but a 40kg weight that he's been carrying around in the back flies out on to the ground where it shatters into four pieces. The assistant is angry and upset. Even though the accident wasn't his fault the merchant will charge him for the damage and deduct the money from his wages.

Then suddenly, being a maths genius, he has an idea, an idea that illustrates beautifully the concept of optimal categorization that we've been thinking about recently: the judicious division of a disparate and diverse world into a varying number of functionally 'optimal' categories depending on context and the needs of the situation at hand. On weighing the four fragmented pieces of his 40kg cargo the assistant quickly determines that with the aid of a balance scale with left and right pans he can measure any weight of goods between 1kg and 40kg.

Question: what are the weights of the four fragments?

Now, there exists an exhaustive mathematical proof behind the solution to Bachet's poser involving brackets, hieroglyphics, various combinations of military-grade alphanumerics and some serious Greek letters. So I suggest we just cut to the chase: 1kg, 3kg, 9kg, and 27kg.

Here, using some random examples, is how it works.

DESIRED AMOUNT OF SPICE	WEIGHTS IN LEFT PAN	WEIGHTS (PLUS SPICE) IN RIGHT PAN
2kg	3kg	1kg
5kg	9kg	4 (3 + 1)kg
14kg	27kg	13 (9 + 3 + 1)kg
20kg	30 (27 + 3)kg	10 (9 + 1)kg
25kg	28 (27 + 1)kg	3kg
38kg	39 (27 + 9 + 3)kg	1kg
40kg	40 (27 + 9 + 3 + 1)kg	

Like sock drawers, train speeds and chopped-up bits of circle, the problem of Bachet is a firm favourite among the ranks of high-school maths teachers. And with good reason. In order to solve it you don't need to be a Hawking or an Einstein. You just need to have the basics: a good, solid grounding in the techniques of algebraic induction and the foundations of a logical mind.

For instance, in order to measure 2kg of spice you *could* use a 2kg weight straight off. On the other hand, however, with a combination of 1kg and 3kg weights respectively you can measure not just 2kg but any weight up to 4kg. Approach the problem from that kind of angle and you're already halfway there.

But there's more to Bachet's conundrum than *just* its value as an intermediate mathematical climbing wall. It's an elegant illustration of a dilemma that's been faced by information architects for centuries, including, most notably, those who design new currencies.

The Optimal Denomination Problem.

How to come up with the minimal number of coins and notes and combinations thereof to enable change to be given most efficiently?

The obvious answer, in light of the puzzle we've just been wrestling with, is to proceed in powers of three (1, 3, 9, 27 are 3^0, 3^1, 3^2, 3^3). In the UK, for example, 1, 3, 9, 27, 81 . . . (pence). Or £1, £3, £9, £27, £81. But there's a problem. Pretty much all currencies these days are decimalized and therefore structurally incompatible with the original Bachet breakdown. Any central bank that starts messing around with its zeros is going to be playing with fiscal fire. Besides, whereas a powers-of-three approach might work for an undiscovered maths genius doling out spices in the dark ages with a set of weights and measures, at the front of the queue at the bar on a Friday night it might prove somewhat less enchanting. Not to mention less efficient.

So, what's the alternative? What is the optimal design of a currency that makes it easiest for its users to part with? Most monetary systems, including the Euro, UK Sterling and US dollar have alighted upon the same structural solution to this enigma.

The 1-2-5 series.

It's an example of what is known as a 'preferred' sequence of numbers. Preferred sequences represent simple orders of integers that can be multiplied by the powers of a convenient base, usually 10, to standardize, simplify and maximize the compatibility between objects, entities and data points within a wide variety of contexts. Of which currency denominations are perhaps the most striking examples.

1, 2, 5, 10, 20, 50 (pence).

£1, £2, £5, £10, £20, £50, £100.

As you will note, the 1-2-5 series isn't a million miles away from Bachet's optimal progression. In fact, it's pretty much identical. The only difference is that whereas in Bachet's sequence the ratio between adjacent integers works out at 1:3, in the 1-2-5 series it equates to 1:2 or 1:2.5. Which means, in terms of currency, that any given denomination should be twice or two and a half times the value of the preceding one, an economic model that facilitates the ease, convenience and accuracy of everyday transactions. Walk into a shop with a £10 note and buy something for £3 and your change *should* come in the form of a £5 note and a £2 coin. Not fourteen 50 pence pieces. Seven £1 coins. Or three £2 coins and a £1 coin. Although we've all been there.

The provenance of the 1-2-5 series provides a salutary demonstration of how just about anything can be simplified should either the need or inclination arise. Even simplicity itself! But alongside the Problem of Bachet it also alludes to a fundamental conundrum faced by each and every one of us during the course of our everyday lives. As we've seen from previous chapters, it isn't just information architects and the designers of new currencies that must grapple with the problem of optimal categorization. It's all of us. From librarians to entomologists. From hoarders to bigots. From politicians to elite sportspeople. Moreover, in everyday life the numbers are generally fuzzier; the solutions not always as crisp and clean-cut as they are with weights and measures.

So how do we know when to stop? How do we know at any given time, within any given context, when we've arrived at the point of 'peak categorization'? When the books on the shelf are optimally spaced and arranged? When we're the optimal distance from the canvas or person or issue in question to appreciate the 'picture' at the level of detail appropriate to the situation at hand? When we're thinking, in other words, in just the right amount of black and white?

It's a question I'd been pondering ever since I'd first encountered the concept of optimal categorization from Mike Anderson over in Perth. And one I didn't have an answer to. I did, however, know a man who would.

Think about it . . . but not too much

Arie Kruglanski is professor of psychology at the University of Maryland in the US.

A Holocaust survivor and founder and co-director of the National Consortium for the Study of Terrorism and Responses to Terrorism (START), he has spent a lifetime behind enemy minds trying to decode the thought processes of violent extremists. His concept of *cognitive closure*, the Nike need to 'just do it', to get the job done and move on, which all of us have to one degree or another, has run him closer to that objective than most.

Sitting in the restaurant of the Four Seasons Hotel in Washington DC, he scratches his head and fiddles with his glasses. 'What,' I've just asked him, 'is the secret to sorting reality? To garnering the right amount of information, in the right number of heaps, in the right size of heap, and with the right amount of intervening space between them?' The answer, it had occurred to me, may well lie in Arie's life's work, in his principle of cognitive closure, and my instinct had proven correct.

He explains to me what he means by it.

'It's basically the desire that each of us has, to greater or lesser extents, to hold on to fixed beliefs in order to keep all of the uncertainty, confusion and ambiguity that life throws at us at bay,' he tells me. 'It evolved through natural selection. Think about it . . . but not too much!

'Thinking is good, right? It's important to give due consideration to different options. But if you have no mechanism to switch that thinking off, to close your mind and reach the point where you say, "Enough is enough, I have now engaged in sufficient deliberation and it's time to make a decision and take action," then what good has that thinking done you? None whatsoever.'

As Arie talks, I'm reminded of Monstromart, a humungous, labyrinthine supermarket in *The Simpsons*, whose slogan, 'Where Shopping is a Baffling Ordeal', says it all. Product choice is unlimited. Shelving stretches right up to the ceiling. Nutmeg comes in 12lb boxes: 'Ooh – that's a good price for 12lbs of nutmeg!' exclaims Marge. And the sign at the express checkout reads '1,000 Items or Less'.

Funny, right? But, like so many things in *The Simpsons*, isn't there a grain of truth in Monstromart, if not the full 12lbs? From shoes to washing powder, to schools, to dating partners . . . isn't it just common sense that the more of something there is, the more we're going to like it?

Well actually, no. The truth is we may *think* we like it. But, in reality, it stresses us out. Too much choice and we freeze, crash and eventually power off. Remember the jam study? The spinning beach ball of over-categorization starts pinwheeling around our brains.

'Don't get me wrong,' Arie continues. 'The ability to deliberate and ponder is one of the truly great perks of having a big brain. Logic, reason, creativity . . . we owe all of our higher faculties to our capacity to compare and contrast. But if there isn't an off switch then where does that leave us? Stuck in a twilight zone of endless, infinite data crunching.

'If we're all research and development and no end product, if we're constantly revving the engine but not actually getting anywhere, then the perks of having a brain the size of ours aren't really perks at all. They're just a drag. Some stupid bug in an otherwise brilliant system. A bug that several million years ago would've quickly seen us off. Would've finished us, in fact, before we'd even started.'

Over the years, Arie has repeatedly demonstrated how the need for cognitive closure can force the brain to quit. It doesn't take much. Elevated levels of noise, time pressure, boredom and fatigue all grab the towel from our fumbling cognitive hands and throw it in early.

The bottom line is: there has to be a bottom line. Our brain needs to have a 'closed' sign in the window. Not all of the time, just some of it, when the place is full and the rooms are all booked out. When we're faced with a riot of itinerant information, when we're checking in mobs of rowdy, rambunctious alternatives, we need to have a notice in reception. One which reads, in big bold letters: 'No vacancies.'

The beauty of Bachet's weights and measures problem is that the solution was written in the cold mathematical stars. The numbers, seemingly, were born to behave that way. But life, as Arie points out, isn't always so disarmingly well-mannered. Rarely does it fall so neatly and conveniently open at the answers page. Because, more often than not, there *is* no answers page. Not one complete with optimal, definitively correct explanations at any rate. Instead, we must muddle through. Consider all the options. Cover all our bases. We must 'weigh up' our choices, 'counterbalance' our judgments, across an entire range of diverse decision-making denominations. What jam should I buy? What candidate should I hire? What school should I send my kids to?

Cognitively, this is costly. Prohibitively so. So a few hundred thousand years ago during the course of our evolutionary history, a cap was imposed on runaway rumination; a failsafe psychological kill switch installed to manage high demands for accuracy on the one hand and to maximize speed and efficiency on the other.

Natural selection isn't stupid. Back in the day it knew perfectly well that if provision wasn't made beforehand our brains would rumble on for ever, continuing to garner interminably more nuanced data on any given problem that arose, dissecting and categorizing its ever diminishing remnants into progressively fractal and increasingly meaningless thought bytes.

'Like the proverbial artist forever sneaking back to the gallery?' I suggest. 'Touching up the unfinished masterpiece?'

'Precisely!' says Arie. 'So it had to do something about it. It had to take action to circumvent the bottleneck. To conjure some means, some reliable, pre-emptive shutdown mechanism to stick a spanner in the works of the spanner being stuck in the works. Some efficient, well-oiled ruse to bring down the curtain should the players get trapped on the stage.'

But what, exactly? Several million years previously, when our prehistoric ancestors were far more likely to have ended up as dinner than as diner, the brain had faced a similar dilemma. Chance upon a snake in the undergrowth or a spider in the corner of the cave and, as we've already seen, how much evidence do you need that it may or may not be harmless?

So natural selection had to step up to the plate. It had to rise to the challenge. Which it did. In fact, the hair-trigger resolution to this everyday primordial imperative produced a slam-dunk adaptation that over time has become synonymous with high-stakes clashes and knife-edge confrontations in pretty much any area of conflict you care to mention. From the battlefield to the playing field. From the canopied jungles of Africa, the Amazon and Australasia to the concrete jungles of Belfast, Baghdad and the Bronx.

Fight or flight.

Might this instinctive, universal response pattern not also have formed the basis of the blueprint to constrain overthinking: the biological prototype for Arie's computational circuit-breaker, the need for 'cognitive closure'? It seems highly unlikely that it wouldn't have had *something* to do with it. Natural selection, after all, is nothing if not frugal. Co-option, the process by which physical structures that arise under one set of conditions can take on different functions and facilitate the transition to new contexts or environments while retaining their original features, is one of evolution's standard operating procedures. If, as Darwin discovered, adaptation generally occurs not through some sudden, one-off, Big Bang moment of mega-metamorphosis but rather through a graduated series of tiny, imperceptible adjustments, then how might 'intermediate' or 'transitional' forms survive, if not by virtue of the fact that features that evolved for one reason at one point in time have the facility, during some later period, to alter their function with minimal modification to their pre-existing structure?*

The lesson from history is clear. Our brain's Force Quit function kept us one step ahead. It was the insistent neural pitchfork in our leisurely cognitive backsides that kept us off the menu. If there languished, in the archives, the original Darwinian design proofs for first-generation 'need for closure' software – fight or flight – then

---

* In *The Origin of Species*, Darwin describes the notion of an already existing structure changing its function as 'an extremely important means of transition' in the course of evolutionary development (Darwin, 1872, p. 175). He goes on to consider traits and features that might, initially at least, serve no apparent purpose due to their incidental emergence as a by-product of evolutionary processes other than natural selection (e.g., the 'complex laws of growth'; the 'mysterious laws of the correlation of parts') but which might, nevertheless, still play an important role in the evolution of those organisms that possess them on account of their potential acquisition of some future function necessary for survival in a new environment: 'But structures thus indirectly gained, although at first of no advantage to a species, may subsequently have been taken advantage of by its modified descendants, under new conditions of life and newly acquired habits.' (Darwin, 1872, p. 186).

at some point or other, any second-generation technology would have built upon the first.

But this isn't to say that we shouldn't still exercise caution, get things in proper perspective, be aware of the problems that are likely to arise when the need for closure shuts up our thinking shops *too* early: biased, blinkered, over-inclusive assumptions; the incapacity, or unwillingness, to see the trees for the wood and discern appropriate levels of subtlety, complexity and diversity; an inability to enumerate a sufficient variety of categories to facilitate the process of informed and rational decision-making. If the need for closure offers us clear and distinct advantages – and, as Arie's research over the past three decades has admirably demonstrated, it undoubtedly does: just ask any trader in any stock market bear pit how much time they have to make up their mind – then it can also have its misuses. Ask any victim of stereotype abuse. Today, in the twenty-first century, bushes don't rustle half as much as they used to and many of the decisions that we make on a daily basis are more complex, less imperative and have wider implications than those made by our prehistoric ancestors.

More often than not all *they* had to draw was the one line. We have to draw quite a few.

And what, I ask Arie, of individual differences? If each of us comes equipped with a Force Quit function, if natural selection has fitted us all as standard with a primeval need for closure to prevent us from cognitively chewing ourselves to death, then is it more a case for some of us than for others? Does each of us differ over how quickly, confidently and decisively we draw our lines? Might there be a continuum of cognitive closure along which everyone has their place?

I recount a story I once heard about Albert Einstein. Apocryphal, perhaps, who knows? One morning, so the story goes, a plumber calls at Einstein's house to undertake some repair work. After greeting him at the door the great man, upon request, gives him a quick tour of the residence. On entering the library, the plumber is dumbfounded by shelf upon shelf of haphazardly arranged tomes. He's incredulous.

'All these books,' he shakes his head. 'It must've taken you so long to read them . . .'

Einstein smiles. 'Most of them I haven't even opened,' he confesses. 'You see, I'm a very slow reader.'

The plumber is mystified. How can such a genius, someone of such peerless intellectual renown, be a slow reader, he wonders? It doesn't make sense. Einstein proceeds to enlighten him.

'Well,' he explains, 'when I do read something I like to make sure that I fully understand it. So I pick one paragraph at a time, read that, close the book and then think about it for a week before opening it again and moving on to the next one.'

Arie gets the picture. 'Like any need – the need for stimulation, for example, or the need to be the centre of attention, or the need to be around other people – the need for cognitive closure varies from person to person,' he says. 'I've actually developed a scale with exactly that in mind, to assess how each of us differs with regard to how closed-minded we are.'

It's tempting to point out that I don't need a questionnaire to tell me that I'm not closed-minded. *I just know.* But instead I take him up on the offer to run me through the form. (If you wish to do the test yourself, my own – considerably shorter – variation of the scale is reproduced in Appendix II on p. 315.) Moreover, it turns out it's not just self-report measures that can quantify our intolerance of uncertainty and ambiguity. In the lab, too, studies examining individual differences in something called 'perceptual perseveration' – the extent to which a particular stimulus or image persists in the 'mind's eye' subsequent to initial exposure – provide additional evidence, to devastating and sobering effect.

Over seventy years ago now, at Berkeley, not long after the war, the Jewish psychologist Else Frenkel-Brunswik conducted an experiment of precisely this type. Now rightfully considered a classic in its field, it lifted the lid on one of the darkest, most sinister personality types in history. The hate-struck mind of the aspiring genocidal despot.

Born originally in Poland in 1908, Frenkel-Brunswik's formative years were peripatetic. In 1914, she and her family had fled across the Austro-Hungarian border to Vienna to escape the pogrom; and later, in 1938, following the *Anschluss*, she had made her way, like many Jewish émigrés of her generation, to America. That year, 1938, was to prove a turning point in Frenkel-Brunswik's career in a number of ways. While heralding the freedom and promise of the New World – love, financial stability, academic tenure – it also foreshadowed a forebodingly familiar old one in the publication of a text called *Der Gegentypus* (*The Antitype*), a despicable treatise from the pen of the German psychologist Erich Jaensch, an enthusiastic Nazi. Given the grievous events to which history would later attest, this loathsome thesis was to shape Else's thinking profoundly.

The central thesis of *Der Gegentypus* was, in hindsight, precipitously zeitgeisty. And empirically, flat-out wrong. It posited that, depending on an individual's ability to sustain a visual stimulus in their mind's eye once that stimulus had been removed from view, there existed two fundamental, archetypal genera of personality. This dichotomous dyad formed the basis of two opposing poles of a continuum. On the positive end – to Jaensch's warped, toxic and dangerous way of thinking – was the elusive *Eidetiker* (literally 'one who has an eidetic memory') representative of a psyche observed almost exclusively in German purebloods and characterized by traits such as firmness, consistency and regularity. On the other was *Gegentypus*, typified by qualities such as 'lability', individuality and tolerance for ambiguity: a form of 'liberalism', Jaensch contended, associated primarily with Jews and foreigners, and which, to the nationalist way of thinking at the time, posed a substantial threat to the tightly woven fabric of German culture.

But Frenkel-Brunswik, quite literally, saw the world somewhat differently. She took issue with Jaensch's lionization of, as she put it, 'the precise, machine-like, unswervingly unambiguous perceptual reaction' – that is, the *Eidetiker*'s unrivalled ability to accurately preserve an image in their mind's eye following occlusion of the initial stimulus – as the archetype of solidity and constancy and repositioned that archetype closer to the centre of the *Eidetiker–Gegentypus* spectrum. There resided, she maintained, in the apprehension of ambiguity, not weakness, or deficiency, or pathology. But significant psychological value.

The crucial plank of evidence had come in the lab, in the late 1940s, in a study, as I say, now widely regarded as a landmark of its time. Participants were presented with a morphed series of sketches, beginning with a cat and ending with a dog, and comprising a transitional progression of increasingly doglike crossbreeds in between (see Figure 6.1).

As the sequence of figures unfolded one by one, participants were required to perform a simple discrimination task: to indicate the point at which the cat was no longer a cat and had accrued a sufficient number of canine flourishes to be nominally reclassified as a dog.

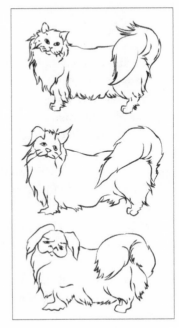

Figure 6.1:
Top – unambiguously feline.
Bottom – unambiguously canine.
Middle – both canine and feline features.

At issue for Frenkel-Brunswik lay the largely involuntary and predominantly unconscious perceptual effects of cognitive flexibility. There would, by the sheer law of averages, always be some participants who would 'see the dog coming' appreciably sooner, or significantly later, than others. But might there be a pattern, she wondered? Natural variation aside, might there be some psychological dimensions, some key personality attributes or overarching character traits that informed these individual differences more than others?

The sole emphasis of the study was on cognitive susceptibility to new information. Or, to look at it the other way, on cognitive resistance to change: that prized, inexorable specialty of the perceptually incorruptible *Eidetiker*. Might those volunteers exhibiting higher levels of prejudice, then, perform differently on the task compared to those more tolerant of conflicting points of view? And even, perhaps, to those more amenable to ambiguity and uncertainty in general? Might the canine metamorphosis happen more quickly in the brains of the moderate than it did in the brains of the radical?

The answer, Frenkel-Brunswik discovered, was yes. And it was a definitive yes at that. Far from being the paragon of mental health and psychological fitness that Jaensch had originally envisioned, the *Eidetiker*, it turned out, was anything but. Not only had those participants who'd exhibited greater levels of prejudice taken longer, on average, to shift their perceptual mindsets from cat to dog, a proportion of them, even right at the very end of the morphed sequence when the transition was undeniable, had resolutely stuck to their guns.

Still they'd seen a Persian instead of a Pekingese!

The lessons from Frenkel-Brunswik's experiment were twofold. One, tolerance for ambiguity fell neatly on a spectrum along which each of us has our place. Two, move on up towards the 'intolerant' end of that spectrum and one can start to lose touch with reality.

Black and white is not black and white – it comes in shades of grey

With the aid of a simple scoring key, Arie runs through my responses on his need-for-cognitive-closure test. While I'm no Einstein, I'm not exactly Gestapo material either. But on the question of Einstein, Arie remains circumspect. Even if the plumber story were true, he points out, it's possible that the great man's languorous reading habits alluded not to a low need for closure but rather to a high need for something else. A need often confused with that for closure. But actually quite different from it. The need for *cognitive complexity*.

'Cognitive complexity is all about how we frame the world,' Arie articulates. 'It's about how we use the basic mental structures that we've acquired through prior experience to perceive and respond to our environment. For each set of similar events and encounters, we develop, over time, particular frames that we place around them. These frames – or ways of looking at life – save our brains a considerable amount of time and energy. Because, once they're in place, they allow us to predict and interpret future events and encounters that either slot directly inside them or are a close enough fit to enable our brains to anticipate the outcome.'*

In other words, we don't have to derive every single thing that happens to us from first principles. The formula for dealing with most of them is already there. Inside our heads. Experience affords us perspective on an issue. How we see it. Encode it. *Frame* it. Perspective dictates how we deal with it. How we assess, evaluate, or react to it.

Frames, in other words, are what categories become when they leave home and get a job. Or, to add another layer of loose

---

* For a thumbnail history of the development of the concept of frames from the German Enlightenment philosopher Immanuel Kant to the present day see Appendix III.

organizational metaphor to the mix, if categories are the crates within which we box up the stuff of life, then frames are the tags we slap on the front of those crates. 'Unborn', for example, in the context of human childbirth, may be the category we use for developing in-utero life forms. But 'foetus' or 'baby' are two very different labels we might subsequently stick on the side.

Arie expounds on the way that these frames, or category labels, work: on the crosstalk and interplay between them. 'How each of these frames relates to, overlaps, and influences the others determines cognitive complexity,' he tells me. 'Some folk have brains that are like an Aladdin's cave of frames. Open the door and there are millions of them. And they're all interleaved and intermixed. These are people who have a high need for cognitive complexity. Like Einstein. People who are capable of distinguishing between, and responding to, very subtle differences between angles, approaches, perspectives and interpretations. Often, they will take a long time to think about an issue, to weigh up the pros and cons of a particular idea, to form an opinion or come to a conclusion about something – as the plumber who dropped in on Einstein that day found out to his surprise.

'On the other hand, some people's brains are stocked with significantly fewer frames. And the ones that they do have are arranged individually and in isolation like precious stones in a box. These individuals tend to see things in black and white. They think in terms of absolutes. An argument or a point of view, for example, is either right or wrong. A solution or intervention is either good or bad. These folk, we say, have a low need for cognitive complexity. They evaluate the world on a smaller number of dimensions than those who have a higher need for it.

'The reason that the need for cognitive complexity and the need for cognitive closure are often confused is because both are directly related to the speed at which we make decisions, how quickly we make up our minds. But they are related to this facility in rather different ways. I think it's fair to say that someone who has a combination of a high need for closure and a low need for complexity – in other words, someone who likes to be certain about things but doesn't like to think about them too much – isn't going to lose too much sleep over the decisions that they make. They're going to be pretty cut-and-dried.

'But someone who has a high need for closure and a high need for complexity – and it's not impossible: cognitive complexity may depend on several factors other than the need for closure, such as the enjoyment of thinking, for example, an aesthetic value placed on either complexity or simplicity, or the intellectual capacity to synthesize complex argument – is going to lose plenty! On the one hand they're going to want answers. But on the other they're going to want nuance. They're going to continue to keep on looking for them.'

It's important to make the distinction. Our brains, Arie suggests, are programmed to think in black and white. And have been for millions of years. But black-and-white thinking is a little more complex than it looks. If our need for cognitive closure determines the speed and alacrity at which we draw the *bottom* line then the need for cognitive complexity is all about the way we draw the rest. The lines between categories and frames. Between 'babies' and 'foetuses'.

Between 'us' and 'them'.

All of us come equipped with the standard black-and-white-thinking software because all of us need to draw lines. But how we draw them, why we draw them and when we draw them varies quite considerably.

Some of us draw more lines, quicker lines, and thicker lines, than others.

In the pages that follow, we'll be looking at the implications of these individual differences in black-and-white thinking for how we communicate with each other. Having established the pivotal role played by categorization in our uniquely human ability to reason – to assess diverse and multifaceted relationships between the kaleidoscopic stimuli of an infinitely complex environment – we'll be shifting our attention away from the *origins* of the mind to how we go about *changing* it.

Namely, to the science of persuasion and influence.

Categories are cognitive commodities. They generate business, competition and trade. And wherever there is trade there is movement – of attitudes, allegiances and perspectives in the case of influence – from one place to another. Movement between groups. Between teams. Between organizations. But fundamentally, and most importantly, between *people*.

Herein lies the importance of those frames that Arie was talking about. Those category 'brands'. Those labels on the sides of the boxes. Learn how to affix them efficiently, securely and prominently, learn how to market these ready-to-think ways of looking at the world to whoever is your target audience, and you'll quickly turn into a much more powerful influencer. You'll find yourself becoming considerably more proficient at getting what you want.

A simple example: migrant or a refugee? Much of today's immigration policy is predicated upon the outcome of this binary, black-and-white categorization. Briefly put, a refugee leaves his or her homeland because they have no other option but to do so. A migrant decamps in search of a better life. It's a straight down the line dichotomy; a clear-cut distinction between two different types of migration – forced and unforced – with national policies the world over becoming increasingly attuned to selectively restricting the former while actively resisting the latter.

Or so it would seem on the surface. In practice, however, shoe-horning people into one or other of these two mutually exclusive categories is easier said than done. The diverse influx of people seeking entry into Europe from bombed-out states such as Iraq,

Syria and Afghanistan, or from the impoverished or war-torn regions of eastern and sub-Saharan Africa, do so for a multitude of different reasons many of which stem not from the one, definitive cause but rather from a complex mix of need, desire and circumstance.

This is not to imply that the importance of drawing a line, of making the distinction between those, on the one hand, who are running for their lives and those, on the other, who are merely seeking a better one, should not be acknowledged. Resources, unfortunately, are limited. Even if hope and despair are not. But rather that the two categories that such a line produces – the 'forced/unforced', 'life/better life' dichotomies that our current ideas of population movement reflect – align tangentially at best, and flat-out asymmetrically at worst, with the fuzzier, sketchier picture of reality.

Needless to say, this matters not a jot to unscrupulous politicians. In fact, it suits them down to the ground. Once invoked, the categories are run like dictatorships. If you're a member of one then you can't be a member of the other. Dual citizenship is outlawed. But while 'migrant' and 'refugee' *are* different, the line that separates these two adjacent territories is blurred and hard to make out. The border between them is artificial. Man-made. It is plonked in the middle of a booby-trapped, psychological no-man's-land and patrolled on both sides by well-trained labels and heavily armoured descriptors.

Which means that whoever controls the language wins the argument.

When, in 2018, the anti-immigration lobby in the US began referring to the migrant caravan of Honduran men, women and children fleeing poverty and gang violence in a region home to one of the highest murder rates in the world – some just in shorts and flip-flops – as an 'invading army' they knew exactly what they were doing. They were endeavouring to shore up a *geographical* border, that between the US and Mexico, by usurping control of a *linguistic* one, that between 'migrants' and 'refugees'. If you draw a line and want to make it stick you lash it down with language. Categories are like muscles. They're invisible until they're defined.

The implications for normal day-to-day living are as deep as they are wide. If we didn't have language, quite simply, we wouldn't have anything. Language puts the colour into our lives. Not, as we're about to discover, just in the metaphorical sense. But in the literal sense, too. Colour, as we saw earlier, comes in seven different brands. Red. Orange. Yellow. Green. Blue. Indigo. And violet. And yet as incredible as it may sound, if we suddenly lost the words for any of them – for red, or for blue, or for green, for example – then we'd never see another rainbow. Not one with a full set of shelves anyway. It would be out of stock of whatever corresponding colour term had vanished from our lexicon.

Fact, it is said, can sometimes be stranger than fiction. And the rainbow is a case in point. Without the *vocabulary* for colour we wouldn't have the *colour* of colour.

We wouldn't just think in black and white. We'd *see* in it, too.

CHAPTER 7

The Rainbow That Might Have Been

❁

Before 1802, cirrus, cumulus, and altostratus clouds hadn't been given names. Untitled before 1802, the shapes were present in the sky, ethereal or ephemeral, presumably since the big bang, but un-designated, until they needed to be. Why then? The world hasn't been fully seen, until it is named.

LYNNE TILLMAN

A FEW YEARS AGO, when we first moved to Oxford, my wife and I paid a visit to our local DIY store. We were doing up our bathroom and were after some paint. We knew the colour we wanted – blue – because we'd seen it in a flier the store had put out a couple of weeks earlier. It was on the hull of a boat that a young Nordic family were wheeling out of their immaculate, tree-lined drive by the side of their minimalist, modernist eco lodge.

When we got down to the store I found an assistant and unscrunched the flier from my back pocket. I pointed. 'Got any of that?' I asked.

The assistant took the flier from me, examined it more closely and wandered over to a workstation. 'Not sure,' he muttered, tapping in some numbers. 'Let's have a look.'

My wife and I exchanged glances. The store was massive. There was paint everywhere. How could they not have blue?

'Are you telling me you don't know off the top of your head whether you've got any blue paint in amongst all this lot?' I asked him, gesturing around.

He punched in some more numbers.

'No such thing as "blue" in this job, mate,' he said, his eyes fixed on the screen. 'What you're looking at here' – he prodded the photo – 'is Velvet Breeze.'

I grabbed the flier and stared at it. The hull of the boat was blue. It was BLUE.

'OK,' I said. 'It might be *called* Velvet Breeze. Technically. But basically it's blue, right?'

He took off his brown – better make that Madagascan Mocha – reading glasses and jabbed them at me.

'Listen,' he said. 'I've been in the paint game twenty-five years and you know what? When you've been in paint that long, blue means nothing. I haven't seen blue since 1989. I've seen Aegean Odyssey. Celestial Haze. Cerulean Rhapsody. But blue?' He shook his head. 'Now' – he shoved his glasses back on and swivelled the monitor round so I could see the screen – 'are you sure it's Velvet Breeze you're after? Because there are a lot of other colours you're probably not aware of. Periwinkle Palace, for instance. That's a good alternative. And, bathroom-wise, you can never go wrong with Cumulus Cotton . . .'

Every time I step into my Lapis Rockpool shower I am somehow reminded of how little we actually see of the world. Our brains are bombarded with an infinite stream of sensory, cognitive and emotional stimuli from an environment as rich in information as there are grains of sand on a beach or, present case in point, sub-shades in a colour spectrum. We are assailed every day by ever-increasing nuance and complexity and, given the limitless amount of fine-grade distinctions between stimuli of a certain 'kind' – paint, pasta, jobs, dating partners, special offers – there exists no logical endpoint to the pinwheeling cognitive process that, deep within our brains, conjures from the chaos a clear and structured picture of subjective, coherent 'reality'. For any one of the problems, dilemmas or conundrums that we face we could, in theory, go on and on for ever, gathering ever more detailed intelligence relevant to the decisions that we eventually have to make.

But we don't, do we? Instead, as we've seen, natural selection has equipped us with a Force Quit function that draws a line in the neural sand; that provides us with a sense of sufficient conviction and certainty that removes the need for further deliberation and which prevents our brains from grinding to a ruminative halt. The infinite cascade of never-ending input is curtailed at a critical mass and the brain computes the lowest, *functional* common denominator between the myriad different data points in question. 'Blue' when it comes to paint. 'Migrant' or 'refugee' when it comes to political asylum seekers.

But language has a guilty secret. Words don't just *define* what we see. They also inform and determine it. To demonstrate, two classic experiments provide a perfect illustration – though if subject matter alone is anything to go by you might be forgiven for thinking they've been plucked from the annals of physics, not psychology. One involves temperature. The other, speed.

Both involve language and its fiendishly Machiavellian sleight of mind.

The first study took place at Yale University in the US in 1950 and was all about the way in which we perceive other people, how we form impressions of them. Harold Kelley, the author of the study and a *real* professor, had two groups of students read about two fake professors. One of them was described as 'rather cold'. The other as 'very warm'. 'Both' professors – in reality, the same guy – subsequently entered the classroom and led a twenty-minute seminar with the group they'd been described to. How, Kelley wondered, would the respective cohorts react to him?

The answer, as he suspected, differed significantly. For a start, those expecting Professor Warm were more likely to participate in the discussion compared to those who were expecting Professor Cold. Secondly, they gave him a better write-up, rating him as more humorous, more sociable and more popular. A nicer guy all round.

In fact, later studies have since revealed just how hardwired into our body's central nervous system this hot and cold thing really is. And it goes way beyond words on a page. Research has found that we judge people more positively in warm rooms as opposed to cold, and in all sorts of other situations that we wouldn't normally think about. Like when we're holding a cup of hot coffee versus iced coffee, for instance. We really do, it would seem, quite literally 'warm' to people.

But it's not just that language fixes the way we think. Or influences how we feel. Words can also control what we *see*, what passes in front of our eyes. Nowhere in the history of psychological science has this ever been better demonstrated than in a study carried out in the seventies by the American memory expert Elizabeth Loftus in what is now widely regarded as the inaugural unveiling of one of the most powerful cognitive biases that we know of: the misinformation effect.

The focal point of the study was a video clip of a minor road-traffic accident – a moving car making contact with a stationary car – which Loftus played to two groups of participants. After they'd watched the video, she posed the same question to each group. At what speed did the one car run into the other?

Incredibly, despite both groups having seen exactly the same clip, they gave radically different answers. The response of one group (averaged across its members) was 31.8 mph. Whereas the other group said 40.8mph.*

---

* The actual speed was a mere 12mph.

The variation was explained by the subtly different wording Loftus employed in the question. Of one group she asked: 'At what speed did Car 1 *contact* Car 2?' Of the other group she asked: 'At what speed did Car 1 *smash into* Car 2?' That sneaky one-word disparity between the two questions made all the difference to the answers.

Not only that, but those witnesses who'd been asked about the *smash* reported seeing broken glass at the scene of the accident even though, in the original video clip, there was none.

No wonder that 'leading the witness' elicits such urgent and vociferous objections in the courtroom. Our memories are highly malleable. Our brains so easily led. If we can't draw a line without conceiving a pair of labels then we can't conceive a label in the absence of drawing a line.

Linguistic determinism

One dark, wet and bitterly cold December evening I dropped in to the Academy Club in London's West End, a small, upstairs space founded some three decades ago by the late journalist Auberon Waugh as a place where literary types could drink and talk about books. I was meeting a friend, Professor Jules Davidoff, the director of the Centre for Cognition, Computation and Culture at Goldsmiths College, University of London. Jules is one of the world's leading experts on colour perception, more specifically cultural variation in colour perception, and for many years has been trying to figure out why we don't, as he puts it, all see the rainbow in the same way, why certain groups of people in certain parts of the world sometimes see colour differently to how you or I might see it.

The answer, he believes, lies in language and the labels we ascribe to categories. The writer George Orwell once observed (in *Politics and the English Language*, published some thirty years before

Elizabeth Loftus's famous car-crash study first saw the light of day)
that if thought corrupts language then language can also corrupt
thought.* He was referring to linguistic lethargy; how lazy, ready-
made phrases – clichés, or cognitive stop signs such as 'ordinary,
working-class people', or 'the Lord works in mysterious ways' – can
lead to lazy, ready-made thinking.† But when it comes to the world

---

* The nature of the relationship between language and thought has long been
debated by philosophers and psychologists alike. Do thoughts influence language or does language influence thought? Broadly speaking, the debate is
dominated by two opposing paradigms. The first underscores the communica-
tive function of language; language, it is contended, is independent of thought
and is simply a tool that we've evolved over time to express what we think and
feel. The second, in contrast – first propounded by the American linguists
Edward Sapir and Benjamin Lee Whorf in the 1940s and '50s and known today
as the Sapir-Whorf hypothesis or 'linguistic determinism' – suggests that language and thought are *inter*dependent and that language has the power, to
lesser or greater degrees, to shape the human mind.

† Such devices, Orwell observed, often 'sound right' but offer little of value to
independent thinkers and seekers of the truth as their sole purpose is to discour-
age meaningful discussion and critical thinking around any given issue. The
mysterious workings of the Lord, for example, serves as a balm to sufferers of
disappointment on the one hand and a justification to recipients of good fortune
on the other through largely identical means: by, in both cases, shutting down
analysis of how human 'controllables' such as errors of judgment or the application of knowledge might be offset against divine will as contributing factors in
outcomes. Needless to say, such laziness of thought can lead, if unchecked, to
stereotyping – *are* working-class people 'ordinary'? Ordinary in comparison to
whom? What does that 'ordinary' mean, exactly? – and, if instigated by a totalitarian political system, to mind control: as Orwell rather ominously pointed out,
those who cannot think for themselves have their thinking done for them. The
American psychiatrist Robert Jay Lifton writes informatively in this regard. In
his 1961 book *Thought Reform and the Psychology of Totalism: A Study of 'Brainwashing' in China*, he writes: 'The language of the totalist environment is
characterized by the thought-terminating cliché. The most far-reaching and
complex of human problems are compressed into brief, highly reductive,
definitive-sounding phrases, easily memorized and easily expressed. These
become the start and finish of any ideological analysis' (p. 429). Nowhere, perhaps, are Orwell and Lifton's observations better encapsulated than in the 'You
think too much' mantra espoused by the Unification Church founded in South
Korea in the fifties by Sun Myung Moon.

of colour Jules takes things one stage further. Drawing a line between red and orange is fine. But in order to do so you have to know what 'red' and 'orange' look like in the first place. Not just the 'good' reds and the 'good' oranges, as Eleanor Rosch might put it. That's easy. But the 'bad' reds and 'bad' oranges, too. In particular, in and around the border crossing between the two colours, you need to know when a 'bad' red really *is* a bad red and not a 'bad' orange. And vice-versa. Your brain, as Arie Kruglanski would say, needs to get red to force quit on orange and orange to force quit on red. But without a *word* for red and a corresponding word for orange we have a bit of a problem: how do we know what we're 'looking for'? If we don't have a proper description of our suspects then how on earth are we going to pick them out?

Jules gives me a tutorial on cutting-edge colour science. Sir Isaac Newton's discovery, in 1671, that white light passing through a prism decomposes into the seven constituent colours of the visible electromagnetic spectrum remains one of the most remarkable in the history of science. But there is, he suggests, something more remarkable still. Were Newton to have brandished his prism not in his laboratory in Trinity College, Cambridge, but in the middle of the rainforest in Papua New Guinea – and to have grown up not in rural Lincolnshire but as a member of the indigenous, hunter-gatherer Berinmo tribe – then things might have turned out very differently in modern-day physics classrooms. The rainbow's seven constituent colours may well have been perceived to be somewhat at variance with the way we label them today.

To see what he's getting at, let's return to the musings of ancient Greece; to the philosopher Eubulides and his inscrutable Sorites paradox. The problem, you'll recall, revolves around a heap of sand. Or rather when, precisely, it becomes a heap. If the addition of just the one single grain can never transform a non-heap into a heap then how, if we start from scratch, do we ever get there? We all know a pile of sand when we see one. And we all know the difference between a pile and a handful and a smidgen. But, in theory, when one thinks about it purely logically, a heap of sand is an illusion. An impenetrable philosophical enigma.

Now the Sorites paradox can work for anything – it doesn't have to be sand – and colour is no exception. To illustrate, consider the following exercise. Suppose we arrange 10,000 colour swatches in a line ranging from bright red (swatch 1) at one end to bright orange (swatch 10,000) at the other. There is no discernible disparity between any two adjacent swatches – in other words, the change in hue between any one swatch and its immediate neighbour does not exceed the 'just noticeable difference' necessary to detect variation – so every swatch appears exactly the same as the one directly next to it in the series.

This being the case, we can construct the following argument:

Swatch 1 is red.

If swatch 1 is red, then swatch 2 is red.

If swatch 2 is red, then swatch 3 is red.

And so on right up to 10,000 . . . if swatch 9,999 is red then swatch 10,000 is red. But clearly swatch 10,000 *isn't* red. It's orange.

The curse of Sorites strikes again.

Across a number of different cultures and in a series of ingenious experiments Jules, along with colleague Debi Roberson, has shown that distinct anomalies arise when it comes to parsing the colour space. Berinmo speakers, for example, enjoy a somewhat truncated passage to their far-flung rainbow's proverbial pot of gold, traversing just five colour spaces on their polychromatic odyssey. In English, thanks to Newton, we pass through seven. One of the reasons for this is that the Berinmo language doesn't contain separate colour words for the terms 'blue' and 'green' but lumps the two together to form one superordinate category, 'nol'. (For a diagrammatic 'mode map' depicting the precise nature of the differences between the Berinmo and English colour categories, see Appendix IV on p. 320.)

Unique? Not entirely. There are other cultures with equivalent colour naming systems.* But noteworthy? Absolutely, Jules asserts. Because while on the surface this may seem of little importance, as nothing more significant than a segment on *QI*, it actually goes a lot deeper. It has profound implications not just for the Berinmo lexicon but also for native speakers' perception of colours across the blue–green spectrum. That is, for the way that they actually *see*.

Consider, for example, an array of twelve green squares arranged in a circle on a computer screen. Eleven of the squares are identical. But one is infinitesimally different from the rest (we're talking here of a discrepancy of one, maybe two, Pantone shades).† Do you think you could pick out which one it is? If you are like most people who attempt this task, the answer is no. Moreover, it will drive you nuts. Yet native Berinmo speakers perform it with little difficulty.

In contrast, however, were I to remove that imperceptibly different green square from the array and replace it with a royal blue one, then the exercise poses less of a problem. Unless, that is, you're a Berinmo speaker, in which case you'll find it intractable: as vexing or flat-out impossible as you or I would the previous one.

The reason for this, Jules articulates, lies not in physiology but

---

* A number of languages, including Herero (Himba), Korean and Tibetan, have the same word for green and blue – or what linguists refer to as the colexification *grue*. In Tibetan, for example, *sngon po* is the term typically used to describe the colour of both the sky and grass while in Korean *pureu-da* can mean either blue, green or bluish green. Other languages, such as Vietnamese, defer to the sky and the leaves as reference points: speakers, for instance, might say 'blue like the sky' (*xanh da trời*) or 'blue like the leaves' (*xanh lá cây*) to separate the two colours. Similarly, the neighbouring Khmer language's term for blue (*bpoa kiaw*) connotes a sense of both blue and green, while words for green, such as *bpao sloek chek srasa* (literal translation: 'the colour of fresh banana leaves'), follow stricter parameters and do not include blue.

† The Pantone Matching System is the telephone directory for colours. It lists thousands of codes that denote particular hues or shades – a bit like the colour swatches you see in paint catalogues. Next-door neighbours can be very difficult to tell apart: just think how easy it is to get two eleven-digit telephone numbers mixed up if only one digit is different.

in language. The brains of both Berinmo speakers and native English speakers are equipped with exactly the same perceptual neural hardware and are equally able to 'see' the colours that pervade the world around them, but they diverge when it comes to which of those colours they *notice*. In order to notice *anything* in our environment we need to perceive it as different. The object of our perception needs to stand out. But then we need a way to define and identify that difference. To isolate and pinpoint what is unique or distinctive about it.

Language performs that function.

Just *how* it performs it is where Jules's ideas have really taken off. Labels draw lines, he argues. Language circumscribes, certifies and authenticates myriad competing aspects of experience. It both manages, and tracks, expectation. The Berinmo brain, Jules suggests, makes short work of isolating the divergent hue in an almost identical display of green because it's conditioned to do so by its indigenously distinctive linguistic colour categories. It has lots of words for lots of different greens. Its viewfinder, in other words, when it comes to the colour green, is set on extreme close-up. It is not, however, conditioned to discriminate royal blue from green because it has no word for blue. For brains that run on English, in contrast, the conundrum is reversed. We have terrible trouble discriminating between near-identical shades of green. But identifying blue in the midst of a sea of green is a chromatic piece of cake.

Let's consider a more familiar illustration of the principle. Imagine that you're working with a class of thirty school kids who all appear largely similar to one another. None of the thirty really stands out. But then the teacher points to one of them at random and whispers in your ear, 'That's Christopher.' Next time you encounter them you'll be keeping your eye out for Christopher. And, as if by magic, he will be readily identifiable to you. Similarly, imagine that, at the same time that the teacher is pointing out Christopher to you, another teacher takes me aside and points out

Robin. Next time *I* encounter the class it will be Robin who jumps out at me just as Christopher jumps out at you. And just as you won't 'notice' Robin, I won't 'notice' Christopher.

Has anything changed objectively in the environment in relation to its physical structure? No. There in the classroom, concealed among their peers, Robin and Christopher still don't stand out from the crowd. To the casual observer they still appear inconspicuous. But where a change *has* taken place is within the lexical landscape. A change has occurred within my linguistic world and yours. Out there in reality, before they were introduced to us, the categories 'Christopher' and 'Robin' still existed. They were sat at their desks all along. But without the linguistic dog whistles of their names, their ostensibly interchangeable forms just wouldn't come to heel. They failed to answer the call of our attention.

Take, as another example, another Robin. The Robin Red Breast. A cursory look at the robin's red breast reveals a shocking truth. It isn't red at all. It's orange. So why, then, do we have the Robin Red Breast and not the Robin Orange Breast? The answer, once again, can be found in language. The term 'orange' derives from the Sanskrit word *nāraṅga*, pertaining to the fruit. (Yes, the colour was named after the fruit and not the other way around.) This designation subsequently evolved into the Arabic and Persian forms, *nāranj* and *nārang* respectively, and entered Late Middle English in the fourteenth century via the Old French term *pomme d'orenge*. However it wasn't until much later, in 1512, that 'orange' notched up its first recorded appearance in the English lexicon as an officially recognized colour name in its own right. Its progenitor was the compound *geoluhread* – literally 'yellow-red' – an extinct species of archaic orange that for centuries had inhabited the medieval colour lexicon untroubled by semantic predators.

Prior to 1512, then, we had a problem. We had Robin Orange Breasts. But no word for orange. We had red and yellow but orange didn't exist. At least not on British shores. Or rather, it *did* exist but we didn't need to *see* it. Seven, eight, nine hundred years ago, before Tango, budget airlines and mobile phone service providers, what call would there have been for the colour? None. Or very little. Until the fruit turned up, that is.

Yes, oranges are native to China and have been grown there since 2,500 BC. And yes, the Romans imported them during the heady days of Empire. But after the fall of Rome midway through the fifth century the truth is that they became, in Western Europe at least, not so much the forbidden fruit as the forgotten fruit. It was only some four hundred years later when the Moors invaded the Iberian Peninsula, crossing the Straits of Gibraltar from North Africa not just with a cargo of Islam but with some strange, brightly coloured round things, that the fortunes of the orange finally began to change. This time it learned to network and stuck around.

Every label that language confers upon us essentially spells Force Quit. In the electromagnetic spectrum, as we've just seen, 'blue' forces quit on 'green', just as 'orange' forces quit on 'red'. In the political asylum spectrum, 'migrant' forces quit on 'refugee'. And, in the amorphous slew of faces in the classroom, 'Christopher' and 'Robin' force quit on their anonymous peer group. As if by cognitive magic – through experience, expectation, through implicit associations between perception, evaluation and judgment accrued and preserved over time – a label draws a line. And as soon as we have a line, we have a difference.

The American linguist Benjamin Lee Whorf articulated it more succinctly. 'We dissect nature,' he observed back in 1940, 'along the lines laid down by our native language.'

Jules Davidoff smiles.

Which is why, he tells me without the slightest hint of flippancy or satire, if Isaac Newton had set up shop not in the cloistral repose of Trinity College, Cambridge but in a steamy jungle clearing in a distant northern outpost of Papua New Guinea, then Apple and Pride would be a colour or two short of a logo.

Of words and needs

Precisely why the Berinmo have just five colours in their cultural palette can be traced, once again, to our brain's hardwired predilection for optimal categorization: the organization of objects or stimuli into the most efficient number of piles. Think about it. There's not a lot of blue in the jungle, is there? But there sure is plenty of green. So why bother with azure and lapis when you've got fern, forest and a million different offshoots in between? On the other hand, what passes as optimal can differ dramatically. Not just across situations. But across cultures, societies, civilizations . . . even species.

'If a lion could talk,' the Austrian philosopher Ludwig Wittgenstein once wrote, 'we could not understand him.' To this day, no one is entirely sure exactly what Wittgenstein had in mind when he penned this notoriously cryptic observation but one interpretation is obvious. While your average king of the jungle would be perfectly happy to roar on ad infinitum about thousands upon thousands of subtle variations of smell, we humans, with our comparatively inferior hooters, would be hard pressed to make head or tail of it. Our olfactory viewfinders are set much further out. In fact, the phenomenological categories that our brains conjure up from reality not only reflect the structure of the world around us, they also reveal the unique cognitive imperatives of individual societies and cultures, past as well as present. And not just when it comes to colour.

Argentinian gauchos, for instance, have at their disposal over two hundred words to describe the colour of horses yet apportion the world of plants into just four categories: *pasta* (fodder); *paja* (bedding); *cardo* (woody material), and *yuyos* (all other plants). Similarly, although the 'Eskimos-have-hundreds-of-words-for-snow' riff has been repeatedly debunked by exasperated linguists the world over, there is, it turns out, a ring of truth to it. While the Inuit language contains no generic term for water there's certainly no dearth of nomenclature for various kinds of *frozen* water. And then, of course, we have our paint specialists with their lexicon of synonyms for blue and their viewfinders set on arcane, imperceptible difference.

Another case in point, equally curious, involves spatial awareness.

Guugu Yimithirr is an indigenous Australian language – it's where the term 'kangaroo' springs from – that lacks a highly specific, and ubiquitous, category of words: those that depict spatial orientation with respect to bodily position, such as 'left' and 'right', or 'front' and 'back'. Instead, all directions in Guugu Yimithirr are provided on the basis of where the interlocutor is standing in relation to the four points of the compass. Native speakers might refer to 'the mirror at the east end of the dresser', for instance. Or 'the man over your south shoulder'. And, in turn, their recollection of past events will be peppered with references to cardinal directions.

Does this mean that there exists a group of super-orientated individuals in an isolated corner of Queensland who have an enhanced sense of direction compared to the rest of us? Absolutely it does. Studies reveal that native Guugu Yimithirr speakers have what's been described as 'perfect pitch for directions'. No matter where they might be or what they can see, irrespective of whether they're moving or standing still, they always know where north is.

Colour, it's important to note, is a special kind of category. Unlike horses, or plants or ice, or even directions for that matter, colour does not possess enhanced significance for just a handful of isolated cultures. Rather, it constitutes an integral component of all visual perception, and embodies functional signalling properties that code universally for three psychological imperatives: caution (think Stop signs), contingency (think camouflage), and convenience (think sport: ever tried to play snooker with all the balls the same colour or football with both teams wearing identical strips?).

But not all colours are born equal. Some, most notably the 'older' colours, have greater evolutionary swagger than their polychromatic siblings. Track, across any language, current or historical, the order of emergence of the different colours – their birth order, as it were – and a distinct pattern unfolds. Green will not appear in a language before yellow. Yellow will not appear before red. Or red before black and white.* Blue pops up after green.

Why this might be can be traced to both the frequency and the salience of each hue within the ancestral natural environment. The earliest vertebrates, the jawless fish (agnatha) of the Cambrian and Ordovician periods (about 450–550 million years ago), lived in modest, shallow lagoons on the crystalline fringes of the mighty teeming oceans. Their diet relied little on aesthetic preference. They hoovered up sustenance from the murky, muddy substrate where sight was of little consequence. As such, their visual sense, much like that of the amoeba, would have been primarily attuned to monitoring the approach of predators through simple fluctuations in ambient illumination: discerning the movement of a sudden shadow, for example. The only 'colours' of any significance to agnatha would have been light or dark. Black or white. Hide or seek.

Scroll forward to the modern era – two million years ago – and the visual systems of our cave-dwelling ancestors would have been similarly oriented to detecting and responding to rapid and unexpected changes in immediate visibility. As alluded to previously, the uninvited presence of something large, betusked and carnivorous in the mouth of the cave would quite literally have turned the lights out. It would, in a front-door kind of sense as well as in a figurative, metaphorical one, have been an open-and-shut case. A matter of black and white. So, no surprise at all then that these two

---

* It is important to stress that not all languages feature the same number of basic colour terms (see Appendix I, p. 313). In English, for example, there are eleven (black, white, red, yellow, green, blue, brown, pink, orange, grey and purple), in Slavic languages twelve (they contain separate terms for light and dark blue), whereas the Dani tribespeople of Papua New Guinea have only three (black, white and red). However, the number of basic colour terms contained within a language has no bearing at all on the lexical order in which those terms materialize.

colours always turn up first. While colour was still important to our prehistoric ancestors, contrast was even more so.

Lazarus Geiger, nineteenth-century German philosopher and linguist, cast his eye a little further along the wavelengths of the electromagnetic spectrum. Inspired by the then British Chancellor of the Exchequer William Gladstone's chromatic excavation of ancient Greek texts – Gladstone discovered that in Homer's *Odyssey* the word 'black' appears on almost two hundred separate occasions, and 'white' on a hundred, while 'red', 'yellow' and 'green' appear less than ten times each, and 'blue', in fact, not at all (Homer, for example, describes the Aegean in *The Iliad* as 'the wine-dark sea') – Geiger unearthed a similar pattern in a number of other works, including the Icelandic sagas, the Qu'ran, ancient Chinese allegories and an ancient Hebrew version of the Bible.

His conclusion was ingenious. The dominance of black and the relative pre-eminence of red and yellow within the colour federation, Geiger proposed, together with the corresponding primacy of those three descriptors within language, derived from our ancient forebears' primeval assignations with nightfall, dawn and sunrise.*

Of the Hindu *Vedas*, for example, Geiger noted the following:

> These hymns, of more than ten thousand lines, are brimming with descriptions of the heavens. Scarcely any subject is evoked more frequently. The sun and reddening dawn's play of color, day and night, cloud and lightning, the air and ether, all these are unfolded before us, again and again . . .

---

* Other hypotheses abound. Red is the colour of blood; red is the colour of flushed skin and therefore (a) a reliable mood indicator and (b) a proxy for evolutionary fitness via blood-oxygen saturation and healthy pigmentation. Red and yellow are the colours of ripe fruit and an important source of nutrients for our tree-dwelling primate ancestors.

but there is one thing no one would ever learn from these
ancient songs . . . and that is that the sky is blue.

Neuroscientific evidence tends to support such a theory. Research
shows, for example, that the emergent hierarchy of colour terms
across all languages and cultures reveals an exact correlation with
the reactivity of our brain's visual cortex to different frequencies
within the visible spectrum. That is, the stronger our brain reacts
to a colour's frequency – the more salient the wavelength, in other
words – the earlier it appears in language.

This being the case (and familiar, well-worn phrases such as
'code red', 'red flag' and 'red rag to a bull' would seem to corroborate
such findings), is it merely coincidence that human hazard-warning
signs, in pretty much any society or culture, exhibit the same combinations of colours – black, red and yellow – that nature outsources
to advertise dangerous creatures? And that blue, a colour glimpsed
comparatively rarely in nature, isn't only the last basic colour term
to appear in any language, but also exhibits psychophysical effects
that are diametrically opposed to the first to materialize, red?* Far
from it. Warning colouration (or, as ethologists refer to it, *aposema-
tism*) is essentially the opposite of camouflage. Its function is to
make the animal – a wasp or a coral snake, for example – highly
conspicuous to would-be predators so that it is noticed, remem-
bered and then studiously and purposefully avoided.

The strategy works through the principles of classical conditioning.
Potential predators associate the presence of the 'unconditioned'

---

* Studies indicate, for example, that while the colour red tends to increase gal-
vanic skin response (a standard measure of physiological arousal), fire up the
brain's emotion engines and elevate blood pressure, blue seems to do the opposite: pulse rate declines and brain activity in areas associated with emotional
processing and stress reactivity attenuates. Similarly, occupants of blue rooms,
it's been revealed, consistently dial up the thermostat an average four degrees
higher than those in red rooms.

stimulus (warning colouration) with that of the 'conditioned' stimulus – a sting, a bite or a toxin – so that avoidance is learned. On subsequent star-crossed encounters, the appearance alone of the offending colouration is then sufficient to deter attack. The greater the salience of the unconditioned stimulus the more useful and instructive the tutorial: the harder and faster the lesson hits home and the longer it sticks in the mind.

It all comes down to the one simple question that is fast becoming a central and recurring theme of this book. How many colour categories, or categories of anything for that matter, do our worlds *require* us to have? The biography of blue is not well known. This staple of sky and innumerable bathroom interiors is, as Gladstone and Geiger discovered, younger than it looks. Several million years younger, in fact. Neither the ancient Greeks nor the ancient Israelites had a word for it. At a similar loss were the Nordic and Indo-Chinese cultures of the time. They didn't need one. Just as *we* didn't need one for orange until the turn of the sixteenth century. Yes, the sky was blue. But it was neither threatening, nutritious nor lucrative. And besides, was it *really*?*

But the ancient Egyptians were different. Unique among early civilizations in harbouring an independent term for the colour, they had very good reason to do so. There was something exclusive to the culture of the Old Kingdom (2686–2181 BC) that necessitated it. It was nothing of particular ecological or evolutionary significance but a cool-headed commercial proposition, which, four millennia on, unites the likes of Levi's, Cadbury and the New York

---

* Guy Deutscher, author of *Through the Language Glass: Why the World Looks Different in Other Languages*, once tried a little experiment at home. Having been careful to never tell his young daughter the colour of the sky he asked her one day what colour she saw when she looked up. She had no idea. The sky didn't have a colour. Eventually, Deutscher reports, she decided it was white, only later to switch to blue. As we've seen, Lazarus Geiger's analysis of ancient scriptural contemplations of the heavens reveals similar ambiguity over the colour.

Giants with the entrepreneurial Pharaohs: the harvesting, production and large-scale export of 'Egyptian blue'.

Having originally unearthed the gemstone lapis lazuli and used it to adorn the tombs of the ancient monarchs and the eyes of Cleopatra, the Egyptians discovered that relying on lapis alone for its exquisite celestial hue was both restrictive and prohibitively expensive. So they dug out their test tubes and Bunsen burners, pulled on their lab coats and safety goggles, and set about finding an alternative.

They succeeded in spectacular fashion. Heating lime, sand and copper into calcium copper silicate they emerged with a powder the colour of the sky and the sea, a powder unassailably destined to become the world's first synthetic pigment. The pigment was used in paintings and tomb-adornment. And also in the manufacturing of pottery, textiles and jewellery. Of considerable prestige, its use spread steadily throughout the Levantine Crescent, through Egypt, Mesopotamia and Greece to the furthest-flung reaches of the Roman Empire.

As did the name of its opulent, sought-after shade.

In the beginning was the need. And *then* there was the word. In the souks and bazaars of the Levant and ancient Egypt, blue had, quite literally, made a name for itself. And whenever I tell people the story of how it got here they look at me as if I'm joking. As if I'm making the whole thing up.

I'm not.

Hard though it is to believe, language is an hallucinogenic. Not only does it make us see things that *aren't* there – recall the broken glass in the car smash study? – it also makes us see things that *are*.

In 2018 a seemingly innocuous Twitter poll unwittingly sparked a debate that polarized the internet into two bitterly opposing factions. It served up, at a stroke, into the court of public consciousness, a deep philosophical issue that was as vehemently and trenchantly debated as a hotly disputed line call on Championship point at Wimbledon.

What colour are tennis balls?

Answers demonstrated considerable variation. Yellow, green, lime green, chartreuse, hi-vis yellow and fluorescent yellow were among some of the suggestions put forward. Who better, then, to settle the argument once and for all than Roger Federer, arguably the greatest striker of a tennis ball in history?

'Hey, Roger,' hollered a voice in the crowd as the twenty-time Grand Slam winner was greeting fans in Chicago. 'Are tennis balls green or yellow?'

Without batting an eye, Federer stroked what he thought was a clear backhand winner down the line. 'Yellow!' he shot back with a smile.

But far from clinching the point, Federer's response only turned up the heat even higher. We all know what 'yellow' looks like. And we all know what 'green' looks like. But can we pin down the band of the spectrum, draw a precise line in the rainbow, where the transition between the two takes place? Where those grains of yellow turn into a heap of green? Where Green Hill matures into Yellow Mountain? The answer is no. Or, if we can, it will be different tomorrow. And different to the line that somebody else might draw in the lemony-limey ether.

And the reason we're unable to do it? That's simple. It's because we don't need to. Never have. At no point in our evolutionary history have circumstances ever thrown up the necessity to dissect the yellow–green colour space in minute detail, categorizing and cataloguing every nuance, every shade, every hue, with pinpoint, forensic precision.* If they *had* then not only would we have a

---

* Though that may now be changing. A growing number of brands have recently been going through the law courts seeking to buy up small private islands in the global colour ocean by applying for trademark protection for hues that bear an iconic association with their name. Best known, perhaps, in this regard is Cadbury's long-running battle to annex a certain shade of purple, Pantone number 2685C – or 'Dairy Milk' purple – for their own exclusive use. But there are

ready-made answer to the tennis ball conundrum, we'd also be in agreement. Not just over the make of colour: yellow or green. But also over the model: Mojito Moss or Lemonade Lawn.

Instead, our day-to-day viewfinders are adjusted and set not to the quantum, microscopic level of the 10,000,000 colours that are *potentially* visible to the naked eye but rather to the level of 'primary colours' that *are*. The level at which the cost of getting it wrong on the one hand and the effort of getting it right on the other are optimally counterbalanced. The level, to borrow a phrase from the American cognitive scientist Herbert Simon, of 'satisficing'. Of both satisfying and sufficing.

Like the shots or scenes in a movie, there are enough to keep the storyline ticking along. But too few or too many and we literally lose the plot. Yellow and green are black and white from a distance. And that distance is set, that ambit defined, by the hand of natural selection. We have a licence to binarize colour. But only from an optimal range.

---

others. Back in 2004 the mobile network operator Orange lost no sleep over turfing easyJet off its colour patch following concerns that the airline company's use of Pantone 021C was a little too close to its own Pantone trademark, 151C. Similarly, in 2018, in a ruling handed down by the European Court of Justice, the French shoe designer Christian Louboutin successfully secured protection for their signature scarlet-soled stilettos. The trademark, registered in Belgium, the Netherlands and Luxembourg, refers to 'the colour red (Pantone 18 1163 TP) applied to the sole of a shoe'. Or, as it's more colloquially known, Chinese Red. Which presumably means that although rival manufacturers are now no longer able to emblazon the soles of their shoes with that particular colour, they still have the green light to call on neighbouring Pantone shades billeted either side of it in the chromatic halls of residence: 18 1662 (Flame Scarlet), for instance. Or 18 1664 (Fiery Red). If not, and we have Orange vs. easyJet II, then where do we draw the line? When does red become not red enough any more to still be called red? Because unless you're a Martian and have 500 different words for the colour, Chinese Red, Flame Scarlet and Fiery Red are all going to appear . . . just red.

CHAPTER 8

The Frame Game

• •
• •

*It suddenly struck me that that tiny pea, pretty and blue,
was the Earth. I put up my thumb and shut one eye, and my
thumb blotted out the planet Earth. I didn't feel like a giant.
I felt very, very small.*

NEIL ARMSTRONG

'HEY, KEVIN,' THE departmental administrator called out as I
snuck into the Faculty building late one morning. 'How
you doing?'

'I'm *dying*,' I replied. 'You?'

His face was a picture. I think he thought I meant it.

'Well, do you think you could get those project reports to me
before you decide to peg it?' he asked. 'The Exam Board meets on
Friday and the buck stops with me if the marks aren't logged. I'll
send your apologies.'

I raised my hand and headed upstairs to my office. He was never
going to have his own agony column. But then I started ruminating
about what I'd said. Dying. That's a pretty strong word. Did I really
feel that bad? Not exactly, I concluded, on reflection. A bit under
the weather, maybe. But I'd been worse. Whatever the case, I definitely wasn't dying. So why did I tell him I was?

A while back when I was at the University of Cambridge I ran a study in which I monitored the speech patterns of forty under-graduate students for an hour a day for a week. The study was very basic. Each student simply recorded their conversations during a randomly assigned sixty-minute slot for seven consecutive days and then submitted the files for analysis. How much hyperbole did *they* engage in, I wondered?

The results really *were* extraordinary. All of the students resorted to using at least seven 'black' or 'white' descriptors for every hour they were monitored. Awesome, gross, terrified, delirious . . . The super-powered adjectives and phrases just tumbled out of their mouths.

Surprised? *Astounded*, even? You shouldn't be. Try the following exercise to see why. Below, you'll find ten pairs of 'black-and-white' descriptors. These are regular, everyday words that most of us will use all the time. Copy the words out on to a sheet of paper and then next to each pair jot down a single word that accurately describes the grey zone between the two extremes. Some of them are easy. Some a bit more difficult.

Here's an example: 'black' and 'white'. To which the obvious answer is 'grey'. (If you didn't get *that* then maybe skip this section and move on to the next one.)

1. Top and bottom
2. Introvert and extrovert
3. Good and bad
4. Passive and aggressive
5. Large and small
6. Rough and smooth
7. Left and right
8. Awake and asleep
9. Tall and short
10. Happy and depressed

With a bit of luck you should now have ten words written down in front of you. If so, take a moment to quickly cast your eye over them. Do you pick up a common thread? If your responses are anything like those of the Cambridge students I tested, you should do. Might all of your 'halfway-house' words be a tad vanilla?

Let's run through some of the possible answers. Well, if you're not top or bottom you're obviously in the middle. So no prizes for nailing that one. If you're not good or bad then you might be described as mediocre. If you don't take a large or a small then you might fit into a medium. If you're not left or right then politically you might be a moderate. And if you're not tall or short then you might be of average build.

Did you get any of those? I'm sure you got one or two. But let's just pause there for a moment and take a brief look at those five words. Middle. Mediocre. Medium. Moderate. Average. Not exactly inspiring, are they? In fact, in the worlds of PR and advertising they're pretty much taboo.

Another thing. Did you start to run into difficulty on the even-numbered pairs? If so then, again, you're in good company. None of the students I tested could plug the gap between passive and aggressive or between rough and smooth with just the one, single word. Some of them couldn't do it with a hundred. This is worrying. If there exists no expedient, readily available terminology in the English language to adequately describe the grey zone between certain pairs of black-and-white adjectives then, thanks to this linguistic famine, we are being forced to both speak and *think* in polarized, black-and-white terms. With unseen and far-reaching consequences.

Labels, remember, draw lines.

Consider, for instance, the last time you suffered a mildly embarrassing experience only to tell a friend that it was 'hideous', 'horrendous' and that you 'felt like you wanted to die'. Did you really want the ground to open up beneath you and swallow you? Or, actually, was it more just a case of wanting to slope away into a quiet corner and pretend that nothing had happened? Alternatively, think about the last time you went to the cinema and, after the film, started talking about 'the bad guy'. Is any 'bad guy' really totally bad? Surely all of them – from Batman's arch-enemy the Joker to *The Silence of the Lambs*' psychiatrist-turned-serial killer Hannibal Lecter – have some redeeming qualities? In the Joker's case, his chaotic confidence. In Lecter's, his gentlemanly genius.

In March 2020 I put my concerns to the lexicographer, etymologist and author Susie Dent, a familiar face, for the better part of the last few decades, to millions in the UK as the chief incumbent of the Channel 4 show *Countdown*'s Dictionary Corner. We were meant to be meeting for coffee in the café of St Mary the Virgin in Oxford's city centre but due to concerns over Coronavirus the café had closed its doors. Even God was self-isolating. I pick up the phone instead.

'You're the psychologist so you'll know a lot more than me,' Susie cautions, 'but I recently read about a study which showed that extroverts use more extreme vocabulary than introverts. They say "sweltering" rather than "hot", for example. Or, if they were talking about your book, they'd say it was "brilliant" rather than "informative"! Apparently, the reason is that extrovert brains require more cortical stimulation to achieve the same level of arousal as introvert brains and one of the ways they can do that is by getting a bigger hit from the language they use. They make it more extreme.

'Do you think the same thing might be going on on a more general level with language – you know, in society as a whole? There are just so many demands on our attention these days that in order to make ourselves heard above the crackle and hiss of all the linguistic white noise out there, the words we use have to be bigger, louder and . . . I don't know . . . maybe more badass.

'You see it quite a lot in advertising, for instance, where competition for attention is like an Olympic sport. If – once the social-distancing restrictions are lifted, of course – I went out to buy eye shadow or blusher I'd be hard pressed to find good old-fashioned "Damask Rose" or "Fresh Plum" any more. Instead, you've got "Glow Job", "Gash" and "Deep Throat" – brands that scream sex from every rouged pore!'

And the trend isn't just confined to advertising. Our penchant for linguistic super-sizing is, as Susie points out, everywhere. Yesterday's weather forecasts used to tell us it would be cold, wet and blustery. Nowadays, they seethe with meteorological melodrama delivering ominous portents of Frankenstorms and weather bombs in which rain is 'organized' and winds 'bombarding'. Governments no longer employ experts, they employ 'tsars'. And if you're applying for a job as a shelf-stacker or street cleaner then good luck with that. 'Stock replenishment executive' or 'surface technician' more like.

The truth of the matter, when you really start to think about it, is sobering. We fall back on these polarized, black-and-white speech patterns all the time: when we talk about films (mind-blowing), food (divine), holidays (of a lifetime) . . . everything. So naturally, when my colleague asked me how I was, I told him I was dying. I caricatured my physical condition. Did I tell him a deliberate lie? Of course not. Why would I lie about how bad I felt? What I *did* do, on the other hand, completely unconsciously, was resort to dichotomous language. I exaggerated my feelings of general, run-of-the-mill sluggishness. And in so doing – and this is where it starts to get rather strange – quite possibly made myself more delicate, lethargic and enfeebled than in reality I actually was. The words we use don't just influence the way *other people* see things. They also have an influence on the way that *we* see them, too.*

---

* In her work on emotional granularity which we touched upon briefly in Chapter 2, the American psychologist Lisa Feldman Barrett has shown that those who draw on a wider range of vocabulary to express their emotions enjoy a

To illustrate, let me tell you about another simple study I conducted. I began by dividing a bunch of undergraduates into two groups. Each group was then handed a list of ten adjectives that they had to slot into the course of their everyday conversation a total of five times a day for a week. One group were given a list of 'extreme' adjectives such as 'brilliant', 'horrific' and 'hopeless' while the other group got a list of 'middle-ground' descriptors such as 'so-so', 'balanced' and 'regular'.

At the end of the week each participant was presented with an image of a sliding greyscale continuum on a computer screen (the continuum didn't appear on its own but rather amongst a number of other 'decoy' tests) and, using a cursor, they had to indicate on the scale precisely where they thought that it 'entered the black zone' and 'entered the white zone', respectively.

The results themselves were pretty black and white. Those students who'd been using the extreme words all week exhibited a

---

number of surprising health benefits, both physiological and psychological in nature. Not only are they less likely to lose control when they're feeling angry, hit the bottle when they're feeling down, or laugh when they think something is funny – inhibitory behaviours all associated with greater emotion regulation – they are also better able to learn, grow and take the positives from adverse situations and difficult emotional experiences. They visit their doctors less, too. A sideways glance at some emotion-laced terms in other languages lends support to such a notion that words aren't just what we say and how we think – recall, a little earlier, the Guugu Yimithirr-speakers' perfect pitch for directions – but also how we feel. To illustrate, those who embrace the Danish concept of *hygge* – a candlelit, deep-winter cosiness encapsulated by blanket forts, inglenook fireplaces, steaming mugs of hot chocolate, and all things woollen – will be less likely to suffer from Seasonal Affective Disorder. The Danes, who practise the concept all year round – penning handwritten thank-you notes, walking an extra five minutes to the stall selling the *extra*-special coffee beans, and lighting bonfires on the beach – are less likely to suffer from *any* emotional disorder, perennially coming out in surveys as the happiest people on earth. Whether, on the other hand, the German word *Backpfeifengesicht* – commonly translated as 'a face badly in need of a fist' – makes the Germans more aggressive is a study, I would suggest, badly in need of a finding.

significantly lower threshold for both 'black' and 'white' – that is, their demarcations of where the black-and-white zones began were much nearer the central point of the scale – than those whose language had been more moderately and conservatively calibrated.

We *see*, it would appear, what we *say*.

Take a look at that word list again. How often do you use the words 'happy' and 'depressed' during the course of everyday conversation? A lot! You've probably already done so today without even realizing it. And with good reason. Cutting to the chase with anchor words like these, tossing the bric-a-brac of life into cognitive black-and-white bargain bins, makes it easier on both ourselves and on others when they stop and ask how we are.

But in the long run it's storing up trouble. As sure as I'm hunched here over my computer typing in these words, such bad linguistic posture will, somewhere down the line, lead to stresses, strains and, in some cases, even deformity. 'Depressed' itself is a case in point. Due to ubiquitous misuse and terminological wear and tear, the medical designation 'depressed' has, in the mind of the general public, become so synonymous with being fed-up, with being a little bit 'under the weather', that people who really *are* depressed – those who are incapacitated by the desperately debilitating psychological illness that is clinical depression – are often perceived as weak. We've all mourned the death of a loved one, been passed over for promotion or been through the break-up of a relationship. We've all felt a bit low from time to time. A bit, well, depressed. But we bounced back, didn't we? Got through it. Pulled ourselves together. Why, we wonder, misled by the prevalence, the insidious democratization of this one-time clinical descriptor, can't *they*?

'It's the same with terms like "awesome" and "epic",' observes Susie. 'In modern-day usage they've become our multi-purpose go-to words for events and experiences that are anything but awesome and epic. Stuff that's just vaguely useful or moderately handy. It's the same with words like "star" and "hero". We call everyone "stars" or "heroes" these days. People who buy us a coffee, give us a lift home, or don't kick up a fuss when we have to reschedule a meeting.

' "Tragedy" and "disaster" are also good examples. "Can't believe he missed that penalty. What a tragedy!" "Oh no, I've run out of tea. Disaster!" What do we do when we're faced with a real tragedy or when a real disaster strikes? Our vocab aisles are empty because we've panic bought all the überwords.'

'Maybe,' she continues, 'that's why emojis have really taken off in recent years. They're like little language chocolates full of semantic sugar that we pluck from the shelves at times of linguistic . . .'

'. . . crises?' I interject.

Susie laughs.

'Need,' she continues. 'It's a real . . .'

I'm about to do it again but this time she beats me to it.

'Shame,' she admonishes. 'Don't even *think* about calling it a tragedy.'

If we're extending the finger of blame, we can point it at our early ancestors. For our prehistoric forebears, the need to discriminate between the basic properties of primitive artefacts and experiences – light, dark, fast, slow, sharp, blunt – and the need to communicate these differences gave rise, upon the emergence of language, to a raft of black-and-white words. It's a binary vocabulary that, as we've already seen, controls to this day our thoughts and conversations and dominates the pages of pretty much any lexicon or dictionary we might happen to pluck from the shelves. It not only conveys to those around us how we feel – our emotions, desires, intentions and attitudes – but informs and sustains such inherently organic, gratuitously suggestible states.

Its precise origins are, of course, unknown. But one possibility is that it comes from 'embodied cognition', the influence not of the mind on the body but of the body on the mind. Might our ancient

ancestors have begun anthropomorphizing the 'this' and 'that' of their primitive, polemical worlds purely on the basis of their own physiological symmetry . . . eyes, ears, hands, feet? It's an ingenious notion and one not without foundation. Within the field of developmental psychology, for example, the concept of 'two-ness' refers to the natural presence of pairs of elements from which a baby derives a sense of 'self' and 'other' – much as 'decimal' derives from 'ten-ness' (from the Latin *decem*) and finger enumeration.*
Indeed, might the concept of handedness be a special case in point? Might the origins of prejudice, discrimination and stigmatization have descended from the fallout of naturally occurring variance in something as trivial and arbitrary as basic manual dexterity?

Again, it's not out of the question. Our forebears, like us, were predominantly right-handed. Dental records, to consider just one portion of the evidence, show that they favoured their right hands when working and probably also when eating.† If ancient tools were designed with right-handed users in mind then it's not *too*

---

* Remnants of the anthropomorphic origin of counting systems can be found in an array of different languages. In Inuit, for example, the word for 'five' is *talimat* and the word for 'hand' is *talik*; in Guarani (an indigenous language native to Paraguay, north-eastern Argentina, south-eastern Bolivia, and southwestern Brazil) the word *po* translates as both 'five' and 'hand'; and in the Ali language (spoken in the south-west of the Central African Republic), 'five' and 'ten' take the respective forms *moro* and *mbouna*: *moro* being the word for 'hand' and *mbouna* representing a contraction of *moro* ('five') and *bouna*, meaning 'two' (thus 'ten' = 'two hands').

† Wear and tear on fossilized cavemen teeth has been studied by scientists at the University of Kansas and indicates that although our ancient forebears weren't without smarts they were also rather clumsy. The researchers discovered that when our prehistoric ancestors processed animal hides they would anchor one side of the carcass in their mouths while holding a cutting implement in their dominant hands to slice it up. The tell-tale imprimatur of scratch marks on their front incisors provides nailed-down evidence as to which hand they used to stabilize the slabs of meat and which hand wielded the knife. Intriguingly, the incidence of left-handedness in the Neanderthal population bears a striking resemblance to what it is today, around one in ten.

much of a stretch that the difficulties encountered by lefties could easily have drawn social ridicule – much as they did a couple of million years later in the wake of the Industrial Revolution when operating machinery similarly designed for their ostensibly less clumsy and significantly more populous counterparts.

A clue might be found in language. We get the word 'left' from the Anglo-Saxon word *lyft*, meaning 'weak'. And we get the word 'sinister' from the Latin meaning 'left'. And not just when it comes to the eclectic etymology of English. This shadowy association between 'right' and 'goodness' on the one hand and between 'left' and 'ill-fatedness' on the other persists in an alphabet soup of tongues including French, Chinese, Spanish, Italian, German, as well as Nordic and Slavic dialects.

Whatever the roots of othering, of binary cognition and black-and-white thinking, there isn't a thesaurus or phrasebook in the world that doesn't bear the imprint of its opposing linguistic thumbprints. Indeed, so enduring is the innate emotive polarity of our vocabulary that one might be tempted to make the argument, as indeed some quite emphatically have, that our mastery of words didn't merely evolve, as the common-sense position might otherwise appear to suggest, as an aid to communication, as a rudimentary and then increasingly complex means of rendering the transmission of information more accurate, systematic and contextually clear, but also as an instrument of deception, sophistry and subterfuge; as a persuasive device, for those who happen to be good at it, to massage, camouflage and exaggerate the truth to get one over on those more linguistically challenged.

Like many Darwinian notions of psychological function, this 'God's grift' take on the evolution of language is up for theoretical grabs. There is evidence both for and against it. Evolutionary provenance aside, few would deny that language is the key to persuasion; that, in both its written and spoken form, it constitutes, post-consciousness, an indispensable medium of social influence.

There are others out there at our disposal: guns, sex, fear, drugs and revenge, to name but a few. But language has notable advantages. For a start – and this is a big one – it is legal. At least most of the time. Secondly, it is democratic. It is readily available to most. And thirdly, and perhaps most importantly, deployed in the right way, at the right time and by the right individual(s), it has the power to enable genuine, lasting and carefully evaluated change – of mind, of perspective, of attitude – in a way that a gun to the head simply can't.

One of the most potent attributes of language in this regard lies in its facility to conduct the flow of argument from one position or standpoint to an opposing ideological terminal, conferring, in so doing, a different or contrasting perspective in much the same way that metal conducts the flow of electricity in a circuit between two opposing poles. Persevering with this analogy, one might substitute 'electrons' for 'facts' and 'current' for 'line of reasoning', such that language may be seen as enabling consenting changes of heart through the transfer of information from perceived positions of greater knowledge or understanding to those of relatively diminished confidence or certainty. In other words, language affords us the capacity to accrue, arrange, select and present information in such a way as to contest contradictory beliefs. It allows us to construct an alternative, antagonistic and apparently 'superior' reality with which we may challenge, confront and ideally supersede the 'diminished' reality of the individual, or group of individuals, whose minds we are trying to change.

Seeing what we say is one thing, as we discovered in Chapter 7. Getting *others* to see what we say . . . that's persuasion.

But there's more to seeing than quite literally meets the eye, whether we're talking about the material objects that make up our physical environment or the psychological artefacts – the values, beliefs, opinions and ideologies – that make up our social environment. Where we see things *from*, the vantage point from which we observe, has just as much an impact on what our brains perceive as reality and what our minds accept as truth as the actual things themselves.

A few years ago now, I ran my electric-circuit analogy past one of the UK's top QCs in his sumptuous, oak-panelled chambers in London's Lincoln's Inn Fields. He got it, but kicked it up a notch.

'It's true,' he remarked. 'Information does travel around the brain like electricity around a circuit. But one thing you've got to remember is that, just like electricity, it takes the path of least resistance. The best barristers are the ones who can arrange the facts of a case, the pieces of the evidence jigsaw, to create the clearest, most coherent picture in the minds of the jury members. In other words, those who are able to make their version of events easier for the jury to believe than the version presented by their opposite numbers.'

Those, you might say, who are master framers. Who are able to slap those labels on the sides of their neatly wrapped arguments – the data, the details, the nub and the nuance – in a way that doesn't necessarily provide the most accurate portrayal of the events contained within them. But which most readily appeals to the cognitive sentiments and psychological sensibilities of those who are judging the case.

If categories are the products, then frames are the trademarks and brands.

Of course, framing doesn't just happen within the hushed and hallowed confines of a court of law. It happens in all areas of our lives. And for all sorts of reasons.

Several years ago the ex-SAS soldier turned bestselling author Andy McNab and I had a mutual friend we were both a bit concerned about. Our friend ran a newsagents in the Herefordshire countryside north of London, lived in a village five miles or so away from the shop, and caught the bus into work every morning. No Mark Zuckerberg, you might think. But not too shabby either.

Except for one thing. The guy had a problem. A big one. After leaving the British Forces with a stellar career and an exemplary reputation behind him, he couldn't stop drinking. And by mid-morning, seven days a week without fail, he was propped up behind his counter two sheets to the wind, talking bullshit to anyone who'd listen and hurling expletives at anyone who wouldn't.

The locals – when they could get a word in – were beginning to talk themselves. And some had started to walk, making the journey up the road to a neighbouring village where there was another, less bibulous, newsagent. On the financial horizon the storm clouds were beginning to gather. What to do?

Early one afternoon, when we happened to be in the area, Andy and I dropped in on him in the shop and took him for something to eat. Andy had had an idea that at the time I thought was madness but which in fact turned out to be genius.

'Has it ever occurred to you,' he asked the guy, 'that if you start cycling into work in the mornings and back home again at night you can save yourself two bus fares? That works out at about three quid a day. Which, on a Sunday night, adds up to five extra pints.'

Later, in the car, I brought the matter up. 'New one on me,' I said. 'Trying to get someone off the booze by encouraging them to stock up on beer money.'

Andy smiled. 'Mike was always a fit lad,' he said. 'Let's see how this pans out.'

The very next morning Mike cycles into work just as Andy has suggested. And in the evening he cycles back home again. That's three pounds in the kitty, he thinks to himself. A 'free' pint down at the Queen's Head on Sunday. As a visual reminder, he takes three pound coins from his coat pocket and deposits them in an old coffee jar which he places on the kitchen windowsill. He does exactly the same thing the next morning. And the morning after that. And by the end of the week, just as Andy had predicted, he's got twenty-one pound coins sitting there in the jar ready to convert into five additional hard-earned pints. He empties it out. Walks into the bar. And proceeds to enter an advanced state of supreme intoxification.

The next week he follows exactly the same routine. And the same the week after that. And he goes on like this for a while. Until eventually, one Sunday lunchtime, he starts to have second thoughts. After several months of being on the bike he's beginning to feel . . . well, just that little bit more chipper and 'with it'. Not only that but he's noticed he's beginning to *look* better, too. Truth is, the whole cycling into work thing has made him start to re-evaluate his life, face the facts in the cold, hard – and sober – light of day. As Andy had pointed out, he'd always taken great pride in his fitness. Change was in the air. Now he was getting 'back into it' again he was beginning to identify more with the 'old Mike'. The Mike before he'd jacked it in. The Mike before he'd swapped the shots he'd so excelled at in the field for the ones that came in glasses.

Andy was right. From that moment on Mike never looked back. And all because he gave up taking the bus so he could have more money to put behind the bar.

Depends on how you look at it

Andy has always been good at little tricks like that. Not so much shedding new light on things as redirecting the beam so that it comes from a different angle. Not so much rethinking the situation as reframing it. Much as it grieves him to admit it, he didn't actually come up with the technique himself. Instead, this profound influence truth was first identified by Aristotle, the father not just of Western philosophy but also of persuasion science, in his classic work of the fourth century BC, *On Rhetoric*. Sometimes, the most effective, powerful and hassle-free method of changing someone's mind is not to attempt to change the way things *are*. But to change the way they *see* the way things are.

To illustrate, consider the following. An English teacher at an all-boys school walks into a lecture room and, on writing the following sentence up on the board, asks the class to punctuate it: *A woman without her man is nothing.* Meanwhile, down the road at an all-girls school, his friend does exactly the same. When, later, they compare notes they're shocked to discover that the results are completely different.

The boys punctuate the sentence like this: *A woman, without her man, is nothing.* While the girls parse it like this: *A woman: without her, man is nothing.*

The words themselves remain unchanged. Not only are they exactly the same, they also appear in exactly the same order. What *has* changed, however, is the syntax. The way, as it were, in which we 'see' those words. And this simple disparity in arrangement completely alters its meaning.

Aristotle's ancient insight into the fundamental secret of successful influence – that if one can't change the world then the next best thing is to change the way one *sees* it – is now acknowledged by persuasion experts as pure scientific gold. But perhaps our most celebrated nod to the precursory wisdom of the ancients has come not in the form of rhetorical or idiomatic analysis but rather in that of Prospect Theory, a model of probabilistic decision-making under conditions of risk and uncertainty developed in the late seventies by the American cognitive psychologists Daniel Kahneman and Amos Tversky. The theory is disarmingly simple, and yet, on the strength of it, on a snowy December evening back in 2002, almost a quarter of a century after its inception, and six years after Tversky's untimely death, Daniel Kahneman took to the stage of the Stockholm Concert Hall to receive a Nobel Prize.

What Kahneman and Tversky had unearthed in Prospect Theory was metaphysical dynamite: a culturally universal, yet hitherto undiscovered predisposition that had, since the days of our ancient ancestors, framed all manner of choices, decisions and judgments that we make during the course of our normal, everyday lives.

We have a stronger desire to avoid loss than pursue gain.

Or, to put it another way, we like to win. But not as much as we hate to lose.

During their years of collaboration together Kahneman and Tversky must have spent many a happy hour compiling the distinctively ingenious scenarios – some real, some hypothetical – that constitute the building blocks of their extensive body of work. One study which falls into the former category of 'real' Kahneman in fact cooked up with fellow Nobel laureate and behavioural economist Richard Thaler and involved randomly assigning a batch of free coffee mugs to one bunch of students while, at the same time, getting a bunch of their peers to offer to buy the mugs from them. On the surface, one might think, a fairly innocuous paradigm.

However, when, at the conclusion of the study, the pair compared the mean prices tendered by the prospective buyers and the owner/sellers respectively, a remarkable difference emerged. The average asking price quoted by the sellers hovered around $7, whereas those students attempting to acquire the mugs offered, in stark contrast, around $3. *Loss* of the mug loomed larger in the mind of the seller than *gain* in the mind of the buyer.

Arguably among the best known of Kahneman and Tversky's hypothetical scenarios is the so-called 'Asian Disease' problem. The dilemma offers participants a choice between two public-health programmes proposed by the authorities to deal with an epidemic (purportedly originating in Asia) that is threatening the lives of 600 people – prescient, to say the least. The options provided are thus:

> Choose Treatment A and save 200 lives.
> Choose Treatment B and you have a 1/3 chance of saving all 600 people and a 2/3 chance of saving no one.

Which is the best one to go for?

In this version of the dilemma, participants display an overwhelming preference for Treatment A, the programme *guaranteed* to save 200 lives. Typically, around three-quarters of people who are surveyed favour this solution.

But there is another, slightly different variation of the problem, which runs as follows. In the case of Treatment A, 400 people will die. However, in the case of Treatment B, there is a 1/3 chance that no one will die and a 2/3 chance that all 600 will die (see Table 1 below for a summary).

Again, which is the best option?

FRAMING	TREATMENT A	TREATMENT B
Positive	Saves 200 lives	1/3 chance of saving all 600 people; 2/3 chance of saving no one
Negative	400 people will die	1/3 chance of saving all 600 people; 2/3 chance of saving no one

Table 8.1: Positive and negative framing effects (from Kahneman & Tversky, 1981).

In this iteration, exactly as Prospect Theory predicted, Kahneman and Tversky's analysis revealed a completely different pattern of results. When presented with this new pair of alternatives, barely 20 per cent of participants went for Treatment A in spite of the observation that, out of a cohort of 600 patients, the prognoses '200 lives saved' and '400 lives lost' amount to precisely the same outcome.

It was Coffee Mug Wars all over again. 'Loss frames', as decision scientists refer to them, triumph over 'gain frames'. The prospect (hence the name of the theory) of 400 people dying – the *loss* of 400 lives – weighed more heavily on participants' minds than the prospect of 200 people being *saved* contributed to lightening the burden.

So what, exactly, *are* cognitive frames? And how does framing work? Given the widespread acknowledgement of its unparalleled persuasive power where, in practice, does that power actually reside?

Persuasion: The art of saying *what* they think *how* they think it

Last year, I was invited to do a keynote address at an away day for the senior leadership board of a multinational insurance company based in the north-west of England. I was on in the afternoon, the final session of the day. But before me, kicking things off in the morning, was a woman by the name of Claire Smith.

Claire, I would guess, is in her mid-sixties. Slight, willowy and with a boyish shock of wavy grey hair, she looks like a cross between Agatha Christie's Miss Marple and Judi Dench's M in *Skyfall*. Now running a consultancy business specializing in arbitration and risk management, she was, in a former life, a legend in certain circles. Claire describes herself, without a hint of reclame or puffery, as an extreme negotiator, and spent time, amongst other placements, in China (she speaks fluent Mandarin) and Pakistan (when the US was in lockdown following the events of 9/11, she was holed up in Islamabad as political counsellor to the UK and Pakistan governments dealing with high-profile matters of 'mutual international concern').

She has two particular claims to fame. One, she is the only Western female ever to have given herself over to being democratically licked by the people of the Republic of North Korea: she once appeared on a postage stamp there. And two, she's one of only a vanishingly small number of women to have ever negotiated successfully with the Taliban. Vanishing and successful, as Claire points out, constituting two opposing poles of a precipitous influence spectrum.

As we sit having tea and biscuits in the lounge of the golf and spa resort hotel in which the event organizers have put us up – all

teak and teal and unliftable tomes on Eastern European body art –
we get around to talking about the art and science of persuasion.
Claire specializes in the former. I specialize in the latter. She nods
when I tell her about framing and the key role it plays in shaping
the influence process. How what you say is important. But how the
way that you say it is even more important. How it's essential that
your message *be* the part. But even more so that it *looks* it.

Crucial to any successful negotiation, she observes, a bit like
when it comes to giving kids medicine, is to ascertain beforehand
the optimal form that what you are endeavouring to say, or the drug
you are planning to administer, should assume. It's as much about
the sugar as the dose. Extreme negotiation, indeed any negotiation,
she continues, is all about building a relationship with the other
party or parties under difficult, dangerous and sometimes plain
deadly circumstances. Because building that relationship, estab-
lishing such rapport, Claire explains, enables one to explore the
beliefs and the value systems that are fundamentally most important to them and to later frame what one says in sympathetic and
appropriate ways.

In other words, I suggest, you've got to pickpocket the other
party's value system and then sell those values back to them with
your own logo etched on the side?

Her eyes narrow. 'Well,' she hedges, 'I wouldn't put it quite like
that. Certainly not to the Taliban anyway! But I can see where
you're coming from. I prefer to frame it like this. Good negotiators
communicate. Bad negotiators broadcast.'

Claire tells me a story from her time in Pakistan which jumps
straight into pole position as quite possibly *the* best example of
framing that I've ever heard. It's an unparalleled demonstration of
how, with skilful and timely insertion, the right persuasive message prepared and presented in the right psychological frame
can – to revisit our art gallery analogy from Chapter 5 – reel anyone's brain in closer to the canvas for a kindlier, less critical look.

Deep in the mountains on the Pakistan–Afghanistan border, it's just another day. *Azaan*, the Muslim call to prayer, rings out across the forests of spruce, juniper and white pine, and eagles soar among the snow-capped upper corridors of the Great Karakoram. Suddenly, in a cold, candlelit school perched precariously on a ledge in a steep, stone-carved village ... terror. A panic-stricken woman blows breathlessly in through the door and informs the two teachers that a deputation of Taliban, notoriously opposed to the education of women, are heading their way.

Fear grips the classroom of thirty-three girls. The boys are across the valley on the other side of the pass. No one is sure as to the Taliban's mood or intentions. But chances are that they won't be delivering crayons. In fact, it's a pretty safe bet that they'll be hell bent on shutting them down.

Quite possibly shutting them up.

Calmly, gently, the teachers ship the girls back off to their villages. They ascend to the hills in their sky blue *salwar kameezes* to their homes in the open-topped clouds. But rather than retreating to safer ground themselves, the women decide to stay put. And some half an hour later, when a squall of boots descends upon the schoolyard, they serve up basins of rosehip and cardamom tea. And saucers of *barfi* and *laddu*.

They put the commander and his men at ease.

They show them around the classroom and the playground. They open and close the cupboards for them gladly. And they encourage them to look at the educational aids they see there, the work sheets and wall charts and flashcards and storybooks tidied neatly away on the desks and tables and shelves. As they run through the curriculum and their day-to-day activities, they explain to the commander how by learning to read and write the girls will be able to progress their Islamic studies and gain a better understanding of the Qu'ran. And how by learning arithmetic they'll go on to make better wives – driving harder bargains in the marketplace and becoming much less susceptible to inept or unscrupulous stallholders trying to hoodwink or cheat or short-change them.

The impromptu tea-powered open day does the trick. The next morning, bright and early at seven o'clock as the school bell echoes far and wide across the hazy mountain passes of an awakening Hindu Kush, the girls are sitting contentedly at their desks ready to begin their first lesson of the day. They have hope. An education. A future. By selecting the right frame, by appealing to a strain of radical, extremist values deeply embedded within the cultural DNA of militant, fundamentalist Islam, the teachers had fought their corner and saved the day. Against daunting, dangerous, desperately unfavourable odds, they had managed to out-Taliban the Taliban. No sooner had the soldiers made their entrance than the teachers had controlled the 'scene' that lay in wait, permitting the men to see the school not as a stain, or a slur, or an attack on Islamic values. But rather as an intellectual outlet, a philosophical training ground in which to instil, facilitate and promote them.

Ingenious? Yes. Novel and unprecedented? Hardly. Ask any marketing guru, brand manager or advertising exec and they'll tell you the same thing: the impact of such preliminary perceptual hijacking on our brain's decision-making flight paths should not be underestimated. In fact, immediate occupancy of the brain's perspective-controlling cockpit is crucially important in ensuring the safe, smooth and psychologically efficient passage of those we wish to influence from *their* point of view to our own. And such fleeting first impressions don't just apply in our assessment of other people. Research has demonstrated time and again that they hold just as much sway over our evaluation and judgment of data, information and persuasive argumentation as they do over our perception of strangers.

We saw this quite clearly, for example, in Daniel Kahneman and Amos Tversky's 'Asian Disease' paradigm, a study which provides a powerful and revealing insight into the effect of this 'attentional kidnapping' in the realm of epidemiological reasoning. Recall how a subtle shift of frame from 'gain' to 'loss' exerted an abrupt and dramatic influence on which of two treatment programmes participants deemed more appropriate?

Other instances abound. Studies show, for example, that:

> Condoms billed as '90 per cent effective' are regarded as considerably more reliable than those described as having a '10 per cent failure rate'.

> Given a choice between two identically packaged meat products, most of us will select the one labelled '75 per cent lean' over the one labelled '25 per cent fat'.

> Given a choice between accepting a 5 per cent *raise* when inflation is 12 per cent ('gain' frame) and taking a 7 per cent *cut* when inflation is zero ('loss' frame) most of us will go for the raise.

> In soccer, in high-pressure, sudden-death penalty shootouts players typically perform worse when a miss instantly produces a team loss (*c.* 60 per cent success rate) than when scoring ensures a win (*c.* 90 per cent success rate).

Frames are the spin doctors – the publicists, the propagandists, the puppet masters – of the category world. If categories are in the business of lines, then frames lead the market in angles.

In fact, during the Brexit debate in the UK back in 2016, this enhanced susceptibility to loss as opposed to gain comprised a key component of the Vote Leave movement's overarching campaign theme, captured by the slogan 'Take Back Control'. As such, it formed a vital, vote-winning piece of the overall victory jigsaw.

'When I researched opinion on the euro the best slogan we could come up with was "Keep Control",' reflected Dominic Cummings, Vote Leave's campaign director, some six months after the referendum. 'I therefore played with variations of this. A lot of people have given me a lot of credit for coming up with it but all I really did was listen.'

Use of the word 'back' was a stroke of influence genius, however. Loss aversion is one thing. But the chance to even the score, to *eradicate* a loss, quite another. The power of 'Take Back Control' was that, in just three words, it managed to trigger both gain – 'Take Control' – and loss – 'Back' – frames in the one, single, explosive rhetorical salvo. Just one reason, then, why Leave left Remain in the blocks. Indeed, Cummings's comments provide yet another illustration of just how psychoactive words can be, and how frames constructed from these potent mind-altering entities can annex perspective and judgment without us even realizing.

Is abortion, for instance, about a woman's 'right to choose'? Or is it about 'murder'? The pro-choice movement argues the former. The pro-life lobbyists, the latter. Are drugs a 'law and order' problem or a 'public health' problem? How you 'see' it determines how you deal with it.

In the Netherlands, for example, the smoke-filled 'coffee houses' that you find on pretty much every other street corner were first introduced not, as many people believe, as legalized drug dens but rather as part of a treatment and prevention programme to protect cannabis users from exposure to harder drugs. The thinking behind it is logical, liberal and anything *but* black and white. Indiscriminate prohibition creates a binary subculture in which drug-users with vastly different habits and unique, individual case histories are lumped together en masse as one and the same cohort. Which, over time, of course, precipitates the distinct possibility of them actually *becoming* the same cohort. In addition, there was also the concern that saddling young people with criminal records might inadvertently push them a little further along the substance-abuse spectrum than they might otherwise have cared to venture.

In the Philippines, in contrast, we see the opposite. While the president, Rodrigo Duterte, demonstrates a similar reluctance to saddle habitual drug-users with criminal records, he exhibits no such reservations about saddling them with state-sponsored death. On assuming power in 2016 under the banner of 'Courage and Compassion', Duterte not only pledged to exterminate tens of thousands of drug-users – setting up dedicated 'special ops' kill squads with precisely such a purpose in mind – he also entreated voters to do the same. Though it is difficult to put an exact number on the death toll since his inauguration, with no independent statistics available, police claimed that in a period of just over a year 3,400 people had been killed in their operations.

'Hitler massacred three million Jews,' Duterte enumerated at a press conference. 'Now there are three million drug addicts. I'd be happy to slaughter them.'

The irony, of course, is that language is the hardest drug of all. And each and every one of us is hooked.

Attention candy

Frames, however, offer us more than just perspective. In addition to securing us a view on things, they also serve two further psychological functions, basic yet indispensable in equal measure. The first of these relates to the question of emphasis. Frames draw our attention to the details of an issue, to the good points, the bad points, the benefits and the drawbacks at the expense of competing features. They signpost, in other words, what is salient. Or rather, what the manufacturer of the frame wants us to *believe* is salient. Sometimes this works in our interests. In health ads, for instance, or when it comes to campaigns relating to the environment or climate change. But at other times it doesn't and we fall victim to misinformation.

Frames are like magic in Harry Potter. They come in black or white.

To get a quick handle on salience, let's welcome back our English teachers of earlier who have another grammatical workout for us. This time, the task is to insert the word 'only' into the following simple assertion: *She told him that she loved him.*

As it turns out, of the eight possible positions in which the word may be interposed, *all* of the available options enable a syntactically correct rendition of the sentence . . .

ONLY she told him that she loved him.
She ONLY told him that she loved him.
She told ONLY him that she loved him.
She told him ONLY that she loved him.
She told him that ONLY she loved him.
She told him that she ONLY loved him.
She told him that she loved ONLY him.
She told him that she loved him ONLY.

However, what the order of distribution changes, quite dramatically in some cases, is the *meaning* of that sentence. And it does so by virtue of the innate persuasive talent of that little word 'only'. On being introduced to the other seven words in the statement it immediately emerges as compositional top dog, elbowing its way straight to the semantic forefront of the innocent lexical ensemble. As such, our brains are drawn to it like a magnet. Its linguistic star quality attracts our attentional spotlight and wherever it goes, whatever it does, becomes front-page news.

It becomes stylistically, and psychologically, salient.

Back in the early nineties, as part of a series of studies into how we make decisions, the American behavioural scientist Eldar Shafir conducted a truly astounding demonstration of salience manipulation in the lab. Shafir presented participants with a hypothetical legal scenario in which two parents were suing for sole custody of their child. Minimal information on both parents was provided to the participants as follows:

PARENT A	PARENT B
Average income	Above average income
Average health	Minor health problems
Reasonable rapport with child	Very close relationship with child
Works average hours	Lots of work-related travel
Relatively stable social life	Extremely active social life

After reading the descriptions the participants were then divided into two groups, each of which was posed a question. One group was asked: 'To which parent would you *award* sole custody?' The other group was asked: 'To which parent would you *deny* sole custody?'

Big difference, right?

Incredibly, however, despite the diametric disparity between the two questions, both groups favoured Parent B over Parent A. In the first cohort 64 per cent ruled *for* them (i.e. they should be awarded custody). In the second 55 per cent ruled *against* them (i.e. they should be denied custody).

Why?

Again, it all comes down to that word *salience*. The way the question was worded had an immediate impact on the criteria that the participants used when evaluating the parents. It framed, right from the outset, the binary impressions they formed of them. Take a closer look at those descriptions and you'll notice a simple pattern. Parent A is 'average' across the board. Parent B, in contrast, comprises a mix of the positive and not-so-positive qualities that in the ego-fired, zero-sum heat of your everyday custody battle can make the crucial difference as to who comes out on top.

With this, quite literally, in mind, then, those participants who were asked to review the case for *awarding* sole custody naturally homed in on the *positive* caregiving attributes exhibited by each parent. While those on the case for *denial* correspondingly zoomed in on the *negative*. But whatever the brief it actually made little difference. Parent A didn't get a look in. Not because they were bad but because they were boring! The line was drawn by the question, not the answer. And regular, ordinary, average Mom and Dad failed to get across it – failed to attract attention – either way.

By way of another illustration, let's return to the abortion debate we touched upon earlier. Depending on which side of the argument you're on, there are two modes of description you could typically use when setting out your stall. *Foetus* and *unborn baby*. On the surface, these two characterizations may appear interchangeable. But, on closer inspection, they convey very different messages and frame the debate in radically different ways.

Foetus constitutes an anatomical, emotionally decontaminated designation that can be used to delineate the unborn progeny not just of *Homo sapiens* but of any mammalian species. It frames the entity in question in objective, taxonomic, unequivocally biological terms. This 'dehumanization' of our formative months of inchoate life appeals discreetly, whisperingly and unconsciously to a powerful and enduring precedent. Since there already exists within society a widespread consensus that animals may be killed for any number of reasonable and legitimate purposes, it demands less of an ethical leap of faith to concede that there may well comprise certain conditions, of variable gravity and differing psychosocial complexity, that permit the termination of a human 'life' in utero. Through the use of the word 'foetus', then, abortion may be portrayed in a more equitable, more functional, less morally repugnant light.

In contrast, the double whammy 'unborn baby' couldn't be

constructed to emotionalize the practice of abortion any more stren-
uously. It recruits not just one frame, but two, both of them standing
in diametric opposition to the dispassionate, perfunctory counter-
point evoked by its foetal antagonist.

Consider the word 'baby'. Baby, within the context of abortion,
frames the concept of 'progeny' as archetypically human as opposed
to generically mammalian. We don't refer to the offspring of other
species, such as llamas, kangaroos and platypuses, for example, as
babies. Or even, for that matter, as infants. But rather as crias, joeys
and puggles (look at a picture and you'll see why!) respectively.

'Unborn' then brings to the table the distinct but related conno-
tation of 'work in progress'. Not through the effrontery of the hard
sell but rather through the power of suggestion. Quietly, subtly,
indirectly, it offers us a vision of a smooth continuum of linear,
incremental humanness between the uncontested, *post*natal status
of personhood and that of its controversial, *ante*natal complement.

The implication is the polar opposite to that elicited by the use of
the word 'foetus'. Since the taking of human life is deemed acceptable only in rare and exceptional circumstances – war and
self-defence – and since the in-utero condition 'unborn' now becomes
an *extension* of that of its opposite number 'born', a continuous categorical dynamic which twiddles the biological dimmer switch from
soft embryonic shadow to glaring, lung-busting sentience, abortion is
construed as iniquitous. As a shameful, irreversible reprieve from
Life Row.

Just a simple choice of words can make all the difference.

Wild horses

Rolling Stones guitarist Keith Richards has a story about Mick Jagger
and Charlie Watts, the band's drummer. One night, somewhere around
the mid-eighties, Jagger dials the phone in Watts' hotel room. It's the
early hours. He's inebriated. And Watts is in bed. Watts picks up.

'Is my drummer there?' Jagger slurs. There's a brief pause and then the phone goes dead. Jagger thinks nothing of it.

Watts has other ideas. He gets out of bed. Then he has a shave. He puts on a suit, shirt and tie, and freshly polished shoes. And then he goes downstairs. Jagger's there and Watts marches straight up to him.

Bam! He punches him in the face.

'Don't ever call me your drummer again,' he chides. He turns smartly on his George Cleverley heels and clips off back upstairs.

'You're my fucking singer!'

When Mick sobered up the following decade he would've done well to reflect on just how he'd framed that question to Charlie that night. Those words 'my drummer' certainly provided a clear indication as to the lead singer's *perspective* on the band's highly volatile pecking order. They also drew attention to the particular nature of the relationship enjoyed – or rather endured – by the two men. In other words, they made it *salient*. But the real problem for Jagger lay not so much in the realms of perspective and salience but rather with the third and final constituent of framing's key psychoactive corollaries.

Its influence on our *judgment*.

By revealing to the world our perspective on a particular issue, and by manipulating the salience of one or more of its essential working parts, we invite our audience to process the outlook or viewpoint we espouse, and to either defend, rescind or adjust their own conclusion(s) accordingly.

When this goes well, and those with whom we're interacting shift their perspective into a position of greater alignment with our own, we call it persuasion. And framing has done its job. But when it doesn't go so well – when the perspective it imparts and the salient points it highlights are so far removed from the subjective inclinations of those we're trying to influence as to fail to broker any realignment whatsoever and elicit, instead, the completely opposite effect, open and outright antipathy – then sometimes, as we saw with Mick and Charlie, all hell can break loose.

One can't always get what one wants.

Of course, in the context of a drunken argument late at night where the two polar ends of a 'creative differences' spectrum are squaring off against each other across a bleary hotel lobby, no one is going to win. Neither represents the ideal target audience. And both the cognitive and emotional distance between the two contrasting positions constitutes too great a divide to enable even the broadest, outermost ripples of the strongest persuasive arguments to make psychological landfall on the shores of the opposing mindset. Conviction is maintained. Opinions retained. And judgment remains unaffected.

But there are times when the ripples do wash up on shore; when the opposing ideology is not beyond the bounds of persuasive possibility a million psychological miles away on a different attitudinal continent. On occasions such as these it's a different matter entirely – and a mark of the skilled persuader that they cultivate conditions to facilitate maximum proximity. Framing an argument so that it falls within what is known as an audience or individual's *latitude of acceptance* (as opposed to their *latitude of rejection*) is an essential constituent of effective and successful influence.

To motorize the metaphor, think of the brain as an overrun cognitive taxi service picking up various bits and pieces of information from the environment – opinions, slogans, mantras, soundbites, mission statements – and dropping them off at numerous different locations between our ears. Like all taxi firms it will have a call-out zone, a radial compass of a predetermined distance outside of which it won't accept the fare.

This is exactly what happens in persuasion: when something or someone we encounter endeavours to change our mind. A telephone rings in a heaving neural Portakabin deep inside our brain – understaffed, overstretched and increasingly unreliable – and we have a choice. We either log the address and go and pick up their argument. Or we decline the job and take another call.

*

To appreciate the mind-bending properties of framing 'fresh off the page', to quite literally 'see' its influence first hand, we need look no further than the jaw-dropping artifice of some of vision science's most powerful optical illusions and consider, by way of analogy, the magical transformation not of arguments, explanations and reasoning but of figures, objects and landscapes right in front of our eyes. Here, within the context of physical as opposed to cognitive perception, we experience up close how perspective, salience and judgment combine and do their thing. If the aim of successful persuasion is to enable those you're persuading to see and approach things differently to the way they saw them before, then the old adage of a picture painting a thousand words could never be more appropriate.

Take, for example, the line drawing shown below in Figure 8.1a, the Shepard Illusion:

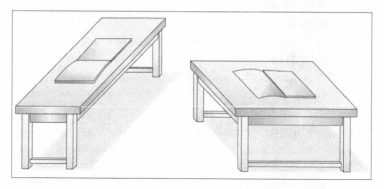

Figure 8.1a: The long and the short of it.

This illusion, named after its originator, the American cognitive scientist Roger Shepard, ranks as one of the most extraordinary sleights of mind ever devised. Even when you know what's going on it's still pretty hard to get your head around it.

So . . . what *is* going on? Well, incredibly, both of the tables in this drawing are the same size. Don't believe it? Here's the proof:

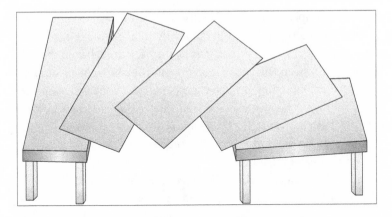

Figure 8.1b: Tables turned.

The power of the illusion resides firstly in the effects of another of its eye-popping counterparts: the Vertical–Horizontal Illusion (see Figure 8.2 below).

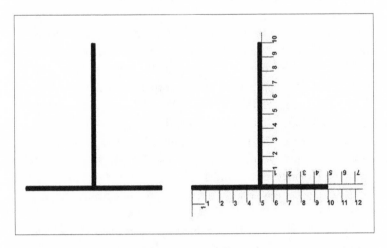

Figure 8.2: The Vertical–Horizontal Illusion.

Stated simply: we tend to amplify and overestimate height. Most people see the vertical stem in the inverted 'T' on the left as being longer than its horizontal base, just as they see the table on the left in Figure 8.1a as being longer than the table on the right on account of its 'taller', more upright, more perpendicular appearance on the page. However, as you can see from the 'T' on the right in Figure 8.2, both lines, just like both tables, are exactly the same length.

Speculation abounds as to why, precisely, we're so susceptible to the Vertical–Horizontal Illusion. One explanation is that it represents an attempt by the brain to compensate for the inequality that exists between the height and the width of our visual field which, it turns out, is actually elongated horizontally in a TV screen configuration to maximize the scope of our left–right peripheral vision.

But there's another reason we fall victim to the Shepard Illusion, which has to do with the tables themselves. Or, more specifically, with the fact that our brain chooses to *interpret* them as tables. Because tables they most certainly are not. In reality, what the illusion *actually* presents us with is a pair of geometric shapes fiendishly arranged at obliquely opposing angles in two-dimensional space. Nothing more. Nothing less. That's it. But there's a problem, of course. These skeletal geometric conformations have appendages. They have 'legs'. Or rather, what we choose to perceive as legs. And when parallelograms sprout legs what do they grow into? Tables.

Exactly the same principles of learning, familiarity and meaning-acquisition apply when, instead of legs, they develop exquisitely configured shaded edges and hang out in the company of other parallelograms, appropriately coloured and strategically aligned, in the middle of the road. On those rare and joyous occasions when such ethereal transformations come to pass, as one did not so long ago in the small fishing town of Ísafjörður, in northwest Iceland, a secondary metamorphosis takes place. The parallelograms turn to stone, hulking great big concrete blocks of stone spookily

Figure 8.3: This 3-D zebra crossing gives pedestrians the feeling of walking on air while simultaneously scaring the hell out of drivers.

suspended in mid-air, with an uncanny ability to stop motorists dead in their tracks (see Figure 8.3 above).

The effect is overwhelming. And the reason for that is simple. Once the brain has made its mind up about what it wants to 'do' with such images it can't help but handle them as three-dimensional objects. Because that, in everyday life, is how our primary visual system has learned to handle *real* tables and *real* concrete blocks. How, in fact, it has learned to handle everything. The brain calls on all its perceptual experience, including, as you'd expect, everything it knows about perspective, to make sense of the ambient trigonometry. To crunch through the lines and the angles to present us, as best it can, with a logical, rational picture we can work with. To conjure, between our ears, the most likely and informative representation of what's 'out there' in front of our eyes.

And so we see what we see. We observe the design in the middle of the road as hovering supernaturally in the ether. And, in

Figure 8.1a, we observe the 'pattern' on the left as receding farther than the 'pattern' on the right. As being longer, in other words.

But go back to Figure 8.1b for a moment and the illusion disappears. As soon as the surfaces of the tables are cognitively unscrewed from their bases and are gradually oriented towards each other at corresponding angles of cumulative geometric proximity, we begin to discern their devilishly disguised dimensions as they really are, in the cold, hard, mathematical light of day.

Because of a *literal* change of frame.

To see how it works, let's return to our three cognitive functions of frames – perspective, salience and judgment – and consider how they might apply. To begin with, removing the legs of the tables completely alters our *perspective* on the image, the primary active ingredient of the illusion. Then, by manipulating the angles of the table tops so that they morph, incrementally, one into the other, we progressively eliminate the difference in spatial orientation between the two surfaces – the *salience* of the perceptual comparison between the vertical and horizontal components of the image – reducing, in the process, the perceived disparity in magnitude between the height and width of the 'tables'. The result is a change in *judgment*. We no longer observe the apparently disproportionate tables as being of unequal size. The illusion vanishes.

Of course, shifting a bit of furniture around is one thing. Shifting minds, attitudes and opinions, quite another. Persuasion, at its core, may well consist of nothing more complicated than the basic psychological manoeuvre of getting people to see things in a different light; of getting them to approach the issue, question or matter at hand from a different, more amenable angle. But even so, does it really make sense to talk of a trigonometry of influence? Is there really any connection between turning the tables literally and turning them metaphorically?

The answer, in short, is yes. More than a connection, in fact; the two might be considered synonymous. Persuasion is psycho-geometry.

It is the careful unscrewing of legs. And the judicious tilting of table tops. Not real table tops. Or real legs. Or the projections of shapes or of recumbent perpendiculars on a page. But those of concepts, arguments, positions and opinions, so that the resulting installation is arranged in such a way that both aspect and composition are agreeable to the eye of the beholder and both structure and orientation accord with the designs of the influencer.

Persuasion is about altering perspective. Persuasion is about dimmer-switching salience. Persuasion, ultimately, is all about the frame. It's about massaging and manipulating judgment.

Do all three and you won't just be *turning* tables. You'll be kicking them over. You'll turn black into white and white into black, trafficking hardcore beliefs and previously inflexible mindsets across armed psychological borders as if they were open and invisible frontiers. As if your words are charmed. And your influence irresistible.

CHAPTER 9

Where There's a Why There's a Way

⁏

*He did what good lawyers always do. He shifted his
argument in the direction his audience was already going.*

JEFFREY TOOBIN

BACK IN 2005 I wrote a book called *Flipnosis: The Art of Split-
Second Persuasion*. The book, as the title suggests, is about
fast, on-the-spot influence as opposed to the more measured machinations of due process and negotiation. It's not just about hitting
the nail on the head, as I put it at the time. But about 'flipping' it
the finger for good measure. The book became quite a hit with
members of the UK military intelligence community, achieving
something of a cult status among their ranks. I'm not surprised.
Given the often acute time constraints inherent to the nature of the
activities with which they are tasked, it was bang on the money.

Not long after *Flipnosis* was published I had the pleasure of
interviewing one of the world's leading academic experts on persuasion, Professor Robert Cialdini, for a BBC radio documentary.
One of the questions I put to him was this: 'Do you think, hypo-
thetically, that anyone can be persuaded to do anything? Or do you
think that there's a limit to the power of social influence?'

Bob hesitated a moment or two before replying. 'I think there probably is a limit,' he said. 'But when you look at tragedies like the Jonestown Massacre it does make you wonder where that limit lies.'*

Some time before, I'd asked another man the same question. My father. Dad wasn't an educated man. He may not have had Bob Cialdini's encyclopaedic knowledge of the principles of social influence, nor his distinguished academic standing. But he was savvy – drop-dead shrewd, as someone once said. Sadly, he's no longer with us. But when he *was* he used to sell things in markets. Most of them dodgy. All of them crap. He *had* to be good to sell them. In fact, let me let you into a little secret. Were it ever to have come down to a one-shot persuasion shootout between Professor Cialdini and my dad as to who could flog the most worthless piece of tat on a market stall in West London on a scrunched-up Sunday morning at the arse end of February, then I'll tell you something for nothing. My money would have been on Dad.

Unlike Bob, he didn't think twice before answering. 'No,' he said. 'I don't believe there is a limit. You've just got to look at what happened in Germany during the war. Provided that you do one thing.'

'What's that?' I asked.

'Make *whoever* is doing *whatever* actually *want* to do it,' he said. 'Persuasion isn't about getting people to do what they don't want to do,' he continued. 'It's about giving them a reason to do what they *do* want to do.'

In other words, whenever you ask someone to do something for you – help you, vote for you, even kill for you – the status of your

* On 18 November 1978, self-styled prophet and leader of the People's Temple the Reverend Jim Jones persuaded a cult of over 900 followers, including some 200 babies and young children essentially murdered by either their parents or cult acolytes, to commit mass suicide by downing a lethal cocktail of cyanide and Kool Aid in a remote commune in the jungle of north-west Guyana.

request depends in no small measure on one rudimentary detail: what's in it for them? It doesn't come any blacker, it doesn't come any whiter, than that. It's the universal category older and wiser than natural selection itself. The category of perceived self-interest.

They say that the great teachers lead by example and if that's the case then Dad was up there with the best of them. Socrates. Aristotle. You name it. Often, he'd try stuff out just to see if he could get away with it. Nine times out of ten he did.

I always remember ambling along a beach in Australia with him one time when we kick past some guy peacefully reading a book under a parasol. He's lying on a towel with his back to us, and behind him, at his feet, is an ice bucket full of beer, the perfect accompaniment to a leisurely afternoon in the sun.

'Fancy a drink, son?' Dad says, as we pull up level with the fella. Assuming he's talking about one of the local bars, I nod. But Dad has other ideas.

'Don't move, mate,' he says, creeping up slowly behind the suddenly paralysed bookworm. 'There's a really big scorpion right between your legs.'

Panic. A month's worth of Ambre Solaire straight down the pan in seconds. The chap goes so white you'd have thought we were in Siberia not Sydney.

'Holy shit!' he screams, frozen to the spot. 'Can you . . . can you kill it . . . ?'

'Well, I'll stand a much better chance if you just shut up and lie still for a moment,' says Dad, silently handing me a couple of chilled Carlsbergs from the ice bucket and trying desperately not to laugh.

'Just keep looking straight ahead and don't move a muscle and we should be all right.'

He waves me away up the beach.

Five minutes later he's still chuckling away.

'See,' he says. 'What have I always told you? Make *them* want what *you* want!'

So maybe my dad was right. But maybe Bob was right, too. Maybe the answer to the question of limits lay somewhere in the middle of what each of them had said. There's a limit to what you can persuade someone to do *against* their will. But if you can get them onside then it's a completely different story.

Dad had mentioned the Second World War with good reason. For him and for other members of his generation the atrocities that had been perpetrated in places like Dachau and Auschwitz had forever come to embody the vilest possible stench of human psychological effluvium. Had he lived to see 9/11 he may well have reconsidered. Did Hani Hanjour, Mohamed Atta, Marwan al-Shehhi and Ziad Jarrah, the respective hijackers and pilots of American Airlines Flights 77, 11, 175 and 93 that within a couple of hours of each other had ploughed into the Pentagon, the North and South Towers of the World Trade Center and a field in Stonycreek Township in Pennsylvania on that tragic, fateful day, actually *want* to complete their heinous, life-shattering missions? Were they so far removed from the natural order of basic human compassion, so diabolically dislocated from the quotidian moral orbits of reason and sensibility as to be acting under their own volition? Or had they acted under duress? It's a moot point. But given the sheer existential enormity of the horrors they brought to bear on that darkest and bleakest of mornings it's hard to believe that they weren't in full concordance with the cataclysmic outcomes of their actions.

The better the reason I give you for doing what *you* want, the greater the likelihood that you end up doing what *I* want. And if I can convince you that it's a good idea to fly a Boeing 767 into one of the world's most iconic buildings claiming thousands of lives in the process – if I can get you, quite literally, to see my black as your white – then I can get you to do pretty much anything.

Between a rock and a hard place

The power of self-interest as an instrument of persuasion and influence isn't trending on social media right now. The secret's been out a while. Take the Old Testament, for example. When, in the Book of Genesis, the serpent prevails upon Eve to partake of the fruit of the Tree of Knowledge, his tactics are strictly old school. He appeals to her vanity. To her desire for position and status:

> Now the serpent was more crafty than any of the wild animals the Lord God had made. He said to the woman, 'Did God really say, "You must not eat from any tree in the garden?" '

> The woman said to the serpent, 'We may eat from the trees in the garden, but God did say, "You must not eat fruit from the tree that is in the middle of the garden, and you must not touch it, or you will die." '

> 'You will certainly not die,' the serpent said to the woman. 'For God knows that when you eat from it your eyes will be opened, and you will be like God, knowing good and evil.'*

We all know how *that* turned out.

But my favourite fable of influence, persuasion and the power of perceived self-interest is a different kind of parable altogether, an ancient folk tale of early eighteenth-century origin that dates all the way back to the days of the French Enlightenment. Commonly known as the 'Parable of Stone Soup', it goes like this.

Early one evening, as the light is beginning to fade, a wise and

---

* Genesis 3: 1–4.

worldly traveller is roaming across the land when he comes upon a small village. As he approaches, the inhabitants of the village immediately become suspicious and run inside, locking their doors and closing their shutters.

The stranger is puzzled.

'Why are you all so frightened?' he asks. 'I'm just a humble traveller looking for a place to rest my weary head and have a hot meal.'

A shutter shoots up and a face appears in the window. 'There's not a scrap of food for miles around,' squawks a mean-looking old woman. 'None of us have eaten for days and our children are all starving. You're better off just staying on the path until you come to the next village on the other side of the mountain.'

The woman points to a large hill a long way off in the distance and glowers at the traveller contemptuously.

He sets down his bundle, peers up at her in the eaves of her crooked little cottage, and gives her a warm, reassuring smile. 'Ah,' he says. 'You have no need to worry. I'm not here to ask anything of you. I have everything I need. In fact, *I* was thinking of making something for *you*. Some of my delicious stone soup!'

From the depths of his cloak he pulls out a sizeable cooking pot. And under the withering gaze of the curmudgeonly old woman fills it with water from a nearby stream. Then, as other shutters open and more faces appear in the windows, he gathers up an armful of brushwood and fallen leaves – sticks, small branches, anything that comes to hand – and proceeds to build a fire, nursing the embers and ensconcing the vessel on top.

And then, with great ceremony, he draws from the stream, as the pot begins to simmer, three smooth, round, medium-sized pebbles and drops them into the water.

'Ahh . . .' he sighs, as the steam rises high into the cool evening air. 'I do like a nice stone stoup! There's nothing quite like it after a hard day on the road.'

By this stage most of the villagers are following his every move, either watching him out of their windows or from the safety of half-opened doorways, their initial suspicions allayed by curiosity. No one has ever heard of stone soup. Could it really be as good as the stranger made out?

A few minutes later, with the sun going down and the pot bubbling away nicely over the fire, he produces a ladle and dips it into the 'broth'.

'Mmmh . . .' he declares heartily, licking his lips. 'Delicious! Of course, yummy though it is, stone soup is always a little better with some cabbage . . .'

He looks around at the growing number of faces in the crowd. The villagers have become quite used to the stranger by now and talk of the soup has started to make them hungry. Sure enough, it isn't long before one of them edges forward with a small cabbage and offers it to him.

'Perfect!' he exclaims, as he chops it up and tosses it into the cauldron. 'Just what I was looking for! Now, you'll never guess what. Several years ago I once had stone soup with a bit of salt beef and it really was the best thing I ever tasted . . .'

Everyone's eyes descend on the village butcher. Five minutes later he returns with a slab of brisket. 'Here,' he says to the traveller, handing him the joint in a folded muslin cloth. 'This should go rather nicely in your soup.'

The stranger takes the beef and adds it to the brew, thanking the butcher fulsomely for his trouble. And so morsel by morsel, potato by carrot, onion by mushroom by cauliflower, the feast grows bigger and bigger.

Until eventually, when there's no more room in the pot, the whole of the village stands gathered around the fire, savouring the aroma with dishes and bowls and soup spoons at the ready.

Persuasion: A unified theory?

It's been fifteen years since I last spoke to Bob Cialdini about persuasion. Now, sitting in a café* in downtown Tempe – he's professor emeritus of psychology and marketing at Arizona State University – we revisit the subject, taking up from where we left off in the BBC studios back in the early noughties.

In 1984, Bob published a bestselling book called *Influence*. It did! In it, he reveals what he describes as the 'six evolutionary principles of compliance', which, having emerged through the exigencies of natural selection, are common to all of us. They are:

Reciprocity – on receiving favours (e.g. gifts, services, invitations) from others we feel obliged to return them.

Scarcity – the less there is of something, the more we want it.

Authority – we defer to those we believe to be credible and knowledgeable.

Consistency – we aim for regularity in, and compatibility between, past, present and future behaviour.

Liking – we prefer to say yes to those we like.

Consensus – we look to the behaviour of others to inform our own actions, particularly under conditions of doubt and uncertainty.

I mention these six principles in *Flipnosis* in the context of my own research into the DNA of social influence. *Flipnosis* is a very different but complementary book to *Influence* in that it focuses on the type of persuasion needed in the *absence* of such core compliance universals. It focuses on spontaneous, single-shot encounters

---

* House of Tricks – you couldn't make it up.

when others do not, for example, owe us a debt of gratitude. Or like us. Or hold power, or authority, or sway over us.

Think barge, not nudge.

As with *Influence*, *Flipnosis* presents a handy set of principles – five of them in all – in the form of the SPICE model, each of which offers a unique contribution to the effectiveness of a persuasive appeal. And each of which, on account of their own evolutionary underpinnings and their role in primeval survival, operates exclusively on the level of self-interest:

> *S-implicity* – our brain has a preference for processing simple over complex information.
>
> *P-erceived Self-Interest* – we are motivated to behave in ways that are personally beneficial to us (or are perceived as being so).
>
> *I-ncongruity* – sudden, unexpected events not only capture the brain's attention but in some cases, as in humour for instance, make us feel good.*

---

* In 2019, scientists at the Max Planck Institute in Germany conducted a study aimed at finding out what, precisely, made a good pop song. The researchers analysed 80,000 different chord progressions from 745 classic US Billboard songs recorded between 1958 and 1991, and used machine learning to ascribe a score to each chord based on how 'surprising' it was compared to the chord preceding it. A representative sample of thirty songs were then selected, stripped of lyrics and melody to obscure the identity of the source track – thus removing the obvious confound of memories associated with different songs – and played to thirty-nine participants. Results showed that when participants were relatively certain as to what chord was coming next and the song unexpectedly took a different turn and surprised them, they found it pleasurable. On the other hand, however, they also felt good if the chord progression was harder to predict but they managed to anticipate it correctly. Follow-up research revealed that this pattern of responses bore a distinct neural signature. Functional magnetic resonance imaging (fMRI) uncovered a significant increase in activity in an area of the participants' brains known as the nucleus accumbens – a reward-related region connected with musical pleasure – both when predictable progressions deviated and when uncertain progressions materialized. 'Songs that we find pleasant strike a good balance

C-onfidence – our brain values conviction and likes to be sure
it's doing the right thing.

E-mpathy – our neural circuits are hardwired to respond
favourably to those we can relate to, and to those who
can relate to *us*.

When it comes to persuasion, SPICE goes back to basics. It's old-school, black-and-white influence in a modern greyscale world that cuts to the chase and constitutes definitive proof that binary can sometimes be best, especially when decisions need to be made quickly and under pressure, and stakes and tensions are high. Perhaps unsurprisingly, one glimpses its spectre in the handiwork of cultists and radicalizers – the simple, the personal, the dramatic, the masterful and the glorious are all popular, go-to narratives for diehard extremists such as the architects of 9/11 – just as much as one does in the exploits of those who defuse.

By way of example, I once heard a story about the boxer Muhammad Ali. Ali was on a plane that was about to take off and was refusing to wear his seatbelt.

'*Please* fasten your seatbelt!' the air hostess pleads with him repeatedly.

But Ali is adamant. 'I'm Superman,' he declares. 'Superman don't need no seatbelt.'

The contender spots her chance. 'Superman don't need no aeroplane!' she shoots back.

Boom! The champ certainly got more than he bargained for

---

between us knowing what is going to happen next and surprising us with something we did not expect,' explains Vincent Cheung, who led the research. The McCartney-written Beatles track of 1968 'Ob-La-Di, Ob-La-Da' – famously decried by John Lennon as 'granny music shit' – topped the researchers' charts hotly pursued by Genesis' 'Invisible Touch', and BJ Thomas' 'Hooked on a Feeling'. Also up there were The Jackson 5's 'I Want You Back', The La's' 'There She Goes', Van Halen's 'When It's Love' and UB40's 'Red Red Wine'.

when he took on this particular challenger. The air hostess's use of SPICE was exemplary. Her riposte was clean, crisp, in Ali's own interest, unexpected, confident, and appealed, no doubt, to his penchant for a knock-out blow. Only this time he was on the receiving end!

Whereas Bob's influence works over time, SPICE works overtime. Which means that in parallel they work *together*. Place both sets of principles side by side, mine and Bob's, and you've got a roadmap for persuasion for near enough any situation you can think of. Plus, a certain degree of overlap. A functional common denominator.

Take a closer look at each of our models of influence in turn and a pattern emerges. There's a unifying thread. Each precept appeals, in some way or other, either to what we like, on the one hand, or to what we don't, on the other. To our own self-interest, in other words. We tend to move towards things that reward us and we steer clear of those that hurt us and cause us pain. In the case of the former, for example, we're suckers for the 'personal touch' in pretty much anything and everything we do; from the targeted use of our first names in political campaign literature to a tap on the arm from the waiter who's taking our order. Therein lies the power of *empathy*. One of *my* principles. In contrast, how many of us relish the feeling of not being able to 'pay someone back' for some favour or kindness that they've done us in the past? Or are enamoured with the nagging suspicion that we might be 'missing out'? *Reciprocity* and *scarcity*, respectively. Two of Bob's principles. Next time you hear the words 'You scratch my back . . .' or see a sign saying 'While stocks last!' you'll know what's going on. No more panic buying of toilet paper next time something like Coronavirus hits.*

---

* In the first few days of March 2020, eyebrows were raised over the sudden rush to go out and buy toilet roll in the wake of the spread of COVID-19. Given the fact the key symptoms of the virus were predominantly non-gastric in nature – a dry, persistent cough and fever – it seemed illogical. The phenomenon was

In other words, put simply, our behavioural patterns and preferences stem from dual motivational origins. For any one course of
action that we might choose over another there are two fundamental super-principles that guide and govern our decision-making.
The anticipation of pleasure. And the circumvention of pain: the
postponement, temporary or otherwise, of the adverse and undesir-
able consequences that will inevitably befall us if we *don't* perform
the action in question; a reprieve from reprisals or punishment as
opposed to the promise of gain or fulfilment.

'Like all species we're motivated by reward,' Bob says. 'By what
makes us feel good. It's very basic stuff. Think about how you might
go about training a dog to, say, obey the command to sit. To start
with, every time you tell it to sit and it does something vaguely sedentary you give it a reward. A treat. You pat it. Or give it a cookie.
Then, as time goes on, you gradually raise the bar. The dog has to
work a bit harder to get a cookie. Vaguely sedentary suddenly
doesn't cut it any more. Instead, it has to assume a posture that is
recognizably sedentary. And so on.

---

certainly no respecter of borders. In the UK, US, and Australia – to name but a
few countries – masses of people were hell bent on stocking up. Why? Dr Steven
Taylor, professor of clinical psychology at the University of British Columbia and
author of the 2019 book *The Psychology of Pandemics* – published in December
of that year: how about that for timing? – posits an interesting theory. 'One thing
that happens during pandemics, when people are threatened with infection, is
that their sensitivity to disgust increases. They are more likely to experience the
emotion of disgust and are motivated to avoid that,' he explained in an interview
with the *Independent*. 'Disgust is like an alarm mechanism that warns you to
avoid some contamination. So if I see a hand railing covered in saliva I'm not
gonna touch it, I'm gonna feel disgust. And that keeps us safe. So there is a very
tight connection between fear of getting infected and disgust. And what better
tool for eliminating disgusting material than toilet paper. I think this is how it
became a conditioned symbol of safety.' Add such fear of infection to the equal
and opposite need to feel prepared – handwashing, though essential, feels a
somewhat desultory gesture in the face of such a potent virus – and over-
shopping, indeed over-compensation of any kind when feeling so powerless and
vulnerable, is inevitable.

'In other words, you reward it incrementally. Until eventually, when it hears the word "Sit!" the dog just does what it's told. It immediately sits down. Just like that. It's known as the principle of positive reinforcement. The oldest persuasion trick in the book. And dog owners are virtuoso exponents of it!'

As are kid owners, too.

I tell Bob about a debate that's been going on in the UK over takeaway tea and coffee cups. With hot beverages becoming ever more readily available in outlets such as supermarkets and petrol stations and the number of coffee shops themselves rising four-fold since the turn of the century, an estimated 2.5 billion cups – all of them potentially recyclable – are being disposed of every year with only 0.25% of the total number produced *actually* being recycled.

The reason behind this woefully inadequate statistic owes more to technological pragmatics than it does to the gremlins of psychological malfunction. Here's how it works. Because the cups have a plastic coating on the inside to enable them to hold liquid and retain heat, they have to be packed off to special decoupling facilities where plastic and paper can be separated. But there's a problem. Only a handful of such premises exist in the UK. So what typically tends to happen is this. While, with the best of intentions, we deposit our cups in the recycling bins available to us, we are, in fact, unwittingly condemning them to an heinous hydraulic death at some anonymous, mixed-waste sorting plant with chutes, loaders and a million mechanical arms. There they are plucked from the conveyors of everyday rubbish and end up not at one of the three dedicated treatment laboratories specifically able to process them, but as landfill.

But what to do about it? Ditch the centralized recycling scheme and introduce 'local' incentives? Some retailers had done exactly that, offering customers anything from £0.25 to £0.50 to slurp their long macchiatos from *reusable* vessels that they'd brought back to the store. No dice. A follow-up government report revealed that just 1 per cent to 2 per cent of all coffees sold were subject to such a discount. And, moreover, that general take-up of the scheme had been minimal. Meanwhile, United Nations sustainability goals concerning waste and pollution continued to pose a challenge. Policymakers began to mumble. If sweeteners weren't the answer then maybe it was time to dish up something sour – a 'latte levy', perhaps? The carrots had had their day. Time to bring out the stick. A charge of £0.25 on disposable cups? Had to be worth a shot.

Bob interjects. The disgruntled policymakers, he suggests, may well have a point; a point backed up by the efforts of behavioural economists.

'The levy makes sound scientific sense,' he says. 'It basically harks back to Daniel Kahneman and Amos Tversky's work on loss aversion back in the late seventies and early eighties. The feeling of *positive* emotion that we experience from gaining a £0.25 discount for returning to a store with a reusable cup is less intense than the feeling of *negative* emotion that we experience when we're hit with a £0.25 levy for not bringing one back.

'Which means that customers should be more motivated to avoid paying the surcharge than they are to receive the discount.' He pauses. 'Again, it highlights the importance of framing in influence. Persuasive communication isn't just about the message. It's also about how that message is put across. It's not just about the present. But about how the present is wrapped.'

A study by Wouter Poortinga, professor of psychology at Cardiff University, demonstrates quite clearly that charging for cups may well be the best way forward. Conducted over a series of trials at a dozen coffee outlets in Cardiff, the results of the study are impressive, revealing an almost 20 per cent hike in hot drinks sales served in reusable cups if the shops provide them for free . . . while at the same time charging for disposable ones.

Plus, as Bob points out, the UK already has a bit of history with this kind of rollout. 'Didn't you guys start charging for plastic bags a while ago?' he asks.

We did. Back in 2015 a plastic bag acquired a charge: £0.05 a pop. And by the end of the first year usage was down dramatically. By more than 80 per cent.

Persuasion isn't about getting people to do what they don't want to do. It's about giving them a reason to do what they *do* want to do.

You know what? I think my dad was on to something there.

No doubt about it

Before Bob and I part company there's something else I'm keen to run by him. The role of reward, self-interest and preference in persuasion has got me thinking about the drivers of influence in a broader sense. In particular, about common influence denominators that may well be operating at an even more fundamental level than those already included in our respective models of behaviour change.

Take another look at the six evolutionary principles of persuasion that comprise Bob's compliance architecture and at the five corresponding components that make up the SPICE taxonomy and a second pattern emerges. One observes that a significant proportion of their combined eleven elements appeals to our hardwired, ancestral intolerance of the unpredictable and indeterminate, and therefore, by implication, to the ensuing need to reduce as much as possible the disorienting, shape-shifting haze of uncertainty and ambiguity.

To turn grey into black and white.

Within Bob's framework, for example, the principles of *authority*, *consistency* and *consensus* each attenuate those uncomfortable feelings of unease, self-doubt and self-recrimination that arise during the course of social interaction. We find safety, in other words, in knowledge (authority), in habit (consistency) and in numbers (consensus).

Similarly, within the architecture of SPICE, the principles of *simplicity*, *confidence* and *incongruity* each touch on the alleviation of confusion, the eradication of hesitation and the maintenance of cognitive consistency, processes that all, in one way or another, speak directly to the diminution of uncertainty . . . albeit ironically, in the case of incongruity, through suitably unorthodox and appropriately incongruous means. Through the use not of evidence or emotional reassurance but through the power of surprise and by actively *disconfirming* expectation.

An ingenious study conducted by Uma Karmarkar, professor of marketing at Harvard Business School, demonstrates what I'm talking about beautifully. Karmarkar began by presenting participants with a positive review of a fictitious Italian eatery, La Scarola. The author of the review awarded the restaurant a highly impressive four stars and cited very good reasons for doing so, flagging up the top-notch food, the first-rate service and the warm, convivial atmosphere. But there was a catch. Karmarkar varied a couple of the food critic's own qualities. First, how certain he was about his review. Secondly, his level of expertise.

In some cases, for instance, the reviewer appeared very certain of his assessment ('I can confidently give La Scarola four stars . . .'). In others, he appeared more circumspect, plagued by self-doubt and conveying hesitation ('I don't have complete confidence in my opinion, but I suppose I would give La Scarola four stars . . .').

A similar ruse was applied when it came to his credentials. To some participants he was described as being a 'nationally renowned food critic and regular contributor to the food and dining section of a major area newspaper', while to others he was portrayed as a 'networks administrator at a nearby community college who keeps a personal Web journal'.

Would either of these variables, Karmarkar wanted to know, have any impact on the participants' own attitudes towards the restaurant, bearing in mind, of course, that they had, in all other respects, read exactly the same review?

The answer was a resounding yes. But, intriguingly, it was neither in isolation that made the difference but a combination of the two together that did the trick. And, moreover, in rather unusual fashion. When participants were led to believe that the critic was an expert they formed more positive impressions towards the establishment on those occasions when he openly expressed *uncertainty* as opposed to those when he said he was certain. In contrast, it was only when the gastronomic status of the reviewer was taken down a peg or two and he was depicted as being less knowledgeable and experienced that certitude rather than doubt began exerting the greater influence on the unwitting would-be diners, rendering their overall judgments of the restaurant correspondingly more in tune with the epicurean accolades they'd read.

And the reason behind this rather odd disparity? Precisely the dynamic that Karmarkar had anticipated. Surprise. Incongruity. And the power of the unexpected.

Consider the manner in which we might envisage most experts presenting themselves. As a rule, we might expect them to be confident in their views. And we might assume non-experts, in general, to be the opposite. We might conceive of them as demonstrating a relative lack of faith in their convictions. And when, in either case, they happen to prove us wrong, when the experts vacillate and the novices display assurance, we can't help but conclude that there's something not quite right. So our brains do a double-take. They blink and pay closer attention. Which leads us, in turn, to examine what they're saying more carefully. And process their opinions more thoughtfully.

Of course, the fact that we have a need for certainty isn't exactly news. As we learned from Arie Kruglanski and his work on cognitive closure in Chapter 6, our intolerance of doubt and the leaden, low-lying brain clouds of decisional ambiguity is of deep and ancient origin. I tell Bob about a study conducted in 2016 by Archy de Berker, a data scientist working out of Montreal, which employed exactly such a paradigm to demonstrate the power of uncertainty in our lives.* More specifically, how even when evaluating the likelihood of *negative* outcomes – such as the chances of receiving a painful electric shock – we prefer surety over ambiguity every time.

The study took the form of a computer game in which participants had to turn over rocks, some of which concealed an unwelcome snake-like surprise. In the event that one did, the volunteers got 'bitten'. Or rather, shocked. It was only mildly uncomfortable, we're not exactly in Death Row territory here. But even so, not exactly pleasant.

As the experiment unfolded, participants were able to deduce which of the rocks were potentially the most hazardous and which were relatively safe. But there was a problem. After a period of time the odds ratios changed unexpectedly. The 'snakes' started moving around. Which generated fluctuating levels of uncertainty and a corresponding variation in participants' physiological stress levels.

Or so one would have thought. In reality, however, it didn't turn out quite like that. De Berker and his team found a definite connection between stress response and levels of uncertainty. That much was clear from the outset. But, curiously, assuming that most of us don't actually look forward to the prospect of an electric shock, the relationship between the two variables wasn't as straightforward as at first it might've appeared. Instead, rather than reporting a general correlation between stress and uncertainty – as uncertainty diminishes and the likelihood of receiving a shock goes up so, too, do stress levels – what the researchers actually discovered was something rather different.

Sure, as the chance of receiving a shock rose from 0 to 50 per cent stress levels went up accordingly. But then the data got a bit weird. As the probability of shock continued to rise from 50 to 100 per cent,

---

* At the time he conducted the study De Berker was based at the Institute of Neurology, University College London.

stress levels, instead of continuing to rise with it, began to come back down. In other words, stress levels were at their highest and lowest respectively when the same held true for uncertainty (i.e., they peaked when the likelihood of receiving a shock stood at 50 per cent and bottomed out when it was at either 0 or 100 per cent).

The conclusion is as clear as it is extraordinary. Such is our dislike of the vague and indeterminate that even a *small* probability of receiving an electric shock weighs heavier on the mind than certain, definitive knowledge that shock is unavoidable.

Even when black and white hurts, even when it causes us actual physical trauma and heralds the approach of inevitable pain and discomfort, we prefer it over grey.

Bob nods in agreement. 'I can believe it,' he says. 'Imagine watching your favourite sports team in a crucial game. When are you going to be more nervous? When the odds are against you and it's a long shot? If the result is a nailed-down certainty and you've got it in the bag? Or when it's the flip of a coin and could go either way?

'For most it's the flip of a coin – when the outcome of the game is hanging in the balance. Same as when you're applying for a job. Or waiting for test results. You'll probably feel more relaxed if you're already pretty sure how the story ends. Or at least if you *think* you are.

'What's interesting about this study, though, is that the outcome is framed in an explicitly negative fashion: an electric shock. Will we get one, or won't we? Yet still we respond in exactly the same way. It just goes to show how powerful a hold uncertainty has over us. And the lengths we're prepared to go to to avoid it.

'Ambiguity reduction is big business these days. Always has been, thinking about weather forecasts and travel timetables and the like. But advances in technology have taken things to a new level. Your iPhone isn't just a phone. It's an uncertainty reduction engine. Even without the apps informing you of the precise location of your taxi and providing you with minute-by-minute updates on your flight status, the very fact that you can make a phone call at all to either impart or receive information remotely, without having to be "there" – wherever "there" might happen to be – marks a huge leap forward from the days when you just had to wait. When you had to leave things, quite literally, "up in the air" until such time as you could physically make the trip and sort things out on arrival.'

Would-be influencers, persuasionistas and thought leaders take note. Corner the certainty market these days and you've got it made. Same as in any day, in fact, as the inventors of gods and the architects of fate and destiny have demonstrated over the years.

The Coronavirus pandemic of 2020 was a long, hard masterclass in freedom and togetherness. But a close third in the hierarchy of lessons learned was that if shutting down schools, queueing up in car parks outside supermarkets, and governments footing wage bills are roads less travelled then grey is the colour least tolerated.

In the UK, we were told that pubs and restaurants were out of bounds. But pubs and restaurants were permitted to remain open for business. We were told that only key workers and those with non-essential jobs who couldn't do them from home were allowed to go in to work so long as they could maintain a safe, two-metre distance in the workplace. But what, precisely, constitutes 'key work'? Doctors, nurses, chemists, delivery drivers and food stores, obviously. But what about bicycle and auto repair outlets like Halfords – designated by the government as an 'essential provider of services' yet nevertheless heavily criticized in the media for implementing a policy of 'partial store coverage' across its 446 shops? What about off-licences – added to the list after supermarkets started running dry? Or social media influencers? In Finland, in 2018, this sector of the workforce was added to the pool of essential service providers following the eerily prescient realization by the communications office of the National Emergency Supply Agency (NESA) that reliance on traditional media would not be enough to reach all four corners of the nation in the event of a major crisis. And what about gun shops in the US – identified as essential businesses by the Trump Administration?

Moreover, what constitutes a safe two-metre distance? Zero employees passing within that radius of each other during the course of a working day? One passing within that distance of just one other on just one occasion? Three passing within that distance of three others on three occasions? The ghost of Sorites rises from the sands once again.

The message is crystal clear. Uncover the means of reducing our ambiguity footprint, come up with a balm for our lingering prehistoric doubt allergy, crack the code of turning grey into hard-edged black and white, and your carriage to power – wherever the road may take you – awaits. Into lockdown. Out of lockdown. Out of the European Union.

CHAPTER 10

Supersuasion

❝❞

If everything isn't black and white, I say, why the hell not?

JOHN WAYNE

THE ROOM WAS spotless, featureless and windowless. The lighting, uniform and cordial. Just off centre, imperceptibly nearer the far wall, stood a table and four chairs. Another chair, unoccupied, sat by the door.

It was eight in the morning. December. Thirteen shopping days to Christmas. Two floors up, at street level, London was heading off to work. Buses were stopping and starting. Shutters were going back. And people were scurrying off to shops and offices with hand-helds, backpacks and polystyrene cups full of strong, steaming coffee.

Down here was a different world, an underworld, where a thick silence clung to the walls and ceilings. The smell was a muggy infusion of coffee, deodorant and chemically clean upholstery. This was a different kind of office with none of the trimmings of the festive city workspaces up above. No cards. No Christmas trees. No pictures of fun-filled nights. Here it was strictly business.

Four people sat around the table. Three men and a woman. I'd met the woman several times before. She was an eminent solicitor who specialized in terror suspect cases. In her mid-thirties, she was immaculately turned out in a graphite, chalk-stripe business suit and had weapons-grade Hillary hair. Next to her sat her client: an Asian man in his early twenties in trainers, Hollister tracksuit bottoms, and a cheap-looking three-quarter-length Puffa jacket. He was a known associate of Abu Hamza al-Masri, the militant Egyptian cleric infamous for preaching hate-filled fundamentalist sermons at his mosque in Finsbury Park, and had been on the UK security service's radar for some time. Recently, something of interest had 'come up' and earlier that morning they'd dropped in on the young man at home at his flat in East London to ask him if he'd mind 'shedding some light' on a couple of matters. Across town. Ever accommodating, they'd even laid on a car.

Now they knew who the brief was, of course, they were screwed. Notoriously censorious and gratuitously canny, there was more chance of Donald Trump turning up in a Just For Men ad than there was of them getting anything remotely useful out of their potential key informant. This solicitor was to candid, unguarded admissions what the Taliban were to aftershave.

I sat on the other side of the table with another psychologist. They'd been plugging away for an hour and were going round in circles. At one point during the questioning, the roles had even reversed and the interviewee, playing the lawyer, had silenced his brief.

'You don't need to answer that,' he'd told her, holding up his hand.

If the stakes hadn't been so high it would have been laughable. That the matter was deadly serious was not overstating the case.

It wasn't long before the interview was terminated. They had nothing to hold him on and even less to charge him with. So that, as the brief concluded curtly, was that. By less than mutual consent it was agreed that a lift back to Dalston constituted an undue abuse of taxpayers' money. The tube, it was politely pointed out, was the way forward.

The solicitor stood up sharply. She zipped up her papers in a black Asprey folder and hovered over the table like a disapproving tiger mother contemptuous of a bum note during a Chopin nocturne. But her client was in less of a hurry. He yawned, looked up at the ceiling, then glanced down at the tape recorder that sat on the table to our left. A small red light indicated that it was off. He sneered.

'You people have no idea about people like Abu Hamza al-Masri,' he said. 'You think he is a madman. A murderer. A criminal. He is not. He is a man of God. A witness to the prophet Muhammad. You denounce suicide bombers as evil, psychotic, brainwashed . . . but at the same time you heap praise on British soldiers in Iraq and Afghanistan who kill innocent Muslims and bomb women and children.

'You are hypocrites. You are snakes. Look into the eyes of the children and ask yourself: who are the heroes and who are the oppressors? Who are the true believers and who are the infidels? Abu Hamza al-Masri is not scared of you. I am not scared of you. We are not afraid to stand up and be counted. We are not afraid to put our necks on the line in the name of Allah.

'Our God is holier than your God.

'Our warriors are mightier than your warriors.

'Our good is deeper than your bad.

'We command our minds to make the biggest sacrifice of all in the service of Allah. We march headlong into death with nothing but steel in our hearts.

'Is there greater love, is there greater generosity we can show to our brothers than that?'

The ancient art and secret science of super-framing

Several years ago, as part of a highly realistic set of counter-terrorism training scenarios, I was asked to sit in and comment on the simulated interviews of suspected Isis associates in what I can only describe as a rather unusual basement in Central London. As an exercise in the study of polemical, antagonistic frames – of how crossed, competing mindsets can generate mutual incomprehension and deep-seated intransigence – it couldn't have been more valuable.

What you have just read is a transcript of one of the encounters I witnessed that day, taken from the notes I wrote up as part of a structured debrief. The role play was second to none. Some of it, it later transpired, wasn't role play at all: those closing comments articulated by an actor had been uttered by a genuine Isis suspect in a real-life interview conducted by the intelligence services. They were sampled because of the extremist, polarizing frame that they presented: a frame offering not only a different perspective from that customarily adopted by upstanding members of the UK security forces but one that spoke to young, disaffected British Muslims whose judgment of society and its abundance of perceived imperfections radicalizers are trying to hijack.

Notice the emphasis on the word 'our'. And how that word is paired firstly with 'God', then with 'warriors', and then with 'good'. Now contrast this with the corresponding emphasis on the word 'your'. And how that is paired with 'bad'. This highly explosive configuration of jet black and pure white frames combining the themes of struggle, moral virtue and a sense of comradeship and solidarity is immensely psychoactive. It's both carefully calculated and meticulously calibrated for maximum persuasive effect, and is regularly deployed not just by political extremists and religious fanatics but by key practitioners in any number of spheres of influence. Law. Advertising. The media. Even kids.

Especially kids!

One Christmas, a couple of years back, I was round at a friend's house having dinner. At the table her ten-year-old son, Josh, took the opportunity to officially announce his forthcoming New Year's resolution. He was going to grow his hair. Mum wasn't too keen and took the opportunity to officially register her considered opposition to the plan.

'I don't think so,' she said. 'Your hair's fine as it is. Short and neat.'

Josh wasn't impressed. 'Spencer's mum is allowing him to grow *his* hair,' he countered. 'When we were in the car the other day, she told his dad that making him cut his hair when he didn't want to was, like, abuse.'

The table fell silent and someone remarked on the broccoli. It later transpired that Spencer's mother was a top human-rights lawyer.

At the time I didn't give Josh's persuasion offensive a second thought. Kids are manipulative, we all know that. But the next morning, over breakfast, I opened up the newspaper and read an opinion piece about the vote in France to ban the niqab, the face veil worn by some Muslim women. The piece focused on three core areas of debate. Issues of security. The preservation of 'French values'. And, of course, the oppression of women. Three binary categories of basic human exigency:

Fight versus Flight. Us versus Them. Right versus Wrong.

The journalist, I reflected, had certainly done her homework. She knew how to get a reaction, which buttons to press to whip up public opinion. Then something occurred to me. Were these not exactly the same existential buttons that Josh had pressed the previous evening over the broccoli? Why bring Spencer's mum into it, unless – *Fight or Flight* – he wanted to make his own mother feel insecure? Or unless, on the other hand – *Us versus Them* – to avail her of 'in-group' values: Spencer's mum and Josh's mum had known each other a long time and agreed on pretty much everything. And what of that word 'abuse'? Did that not connote a distinct sense of the moral? Of *Right versus Wrong*?

Of course, it could've been a coincidence that Josh and this journalist had singled out precisely the same three sets of frames in which to display their persuasion handiwork. But what if it wasn't? What if it emerged that, when it came to getting our way with those around us, there existed a covert, hardwired, optimal number of 'superframes' within which to present one's pitch? A golden configuration of compliance categories within which to slot one's arguments, appeals and requests that brought into the starkest relief those three psychological principles of certainty, closure and self-interest that we earlier identified as key evolutionary cornerstones of social influence?

We encountered this notion of optimal categorization in Chapter 7 with the division of the colour space: how there exist innumerable quasi-identical hues along an infinite Sorites-style continuum (e.g. when does red become yellow?), and yet how our brain perceives just a handful of primary colours. What if such a strategy also holds true for the bending of wills and the changing of hearts and minds? What if the spectrum of all known influence consists of just three primary shades? A 'supersuasion' rainbow?

Let's consider the polarizing rhetoric of Donald Trump. During his successful 2016 US election campaign, precisely what kind of America did Trump pledge to 'make great again'? An America addled with a sense of anger and fear; fear that American values were being unjustly eroded by whoever, at the time, Trump deemed a suitable target. Muslims. Migrants. Mexicans. The 'politically correct'. For America to be 'great again', Trump announced, these people had to be excluded by whatever means possible. Walls. Bans on immigration. Draconian changes to state legislation. Ridicule.

Fight versus Flight. Us versus Them. Right versus Wrong.

More recently, in the face of the Coronavirus pandemic of 2020,

Trump's reference to the strain as a 'foreign virus', his self-laudation of his 'early, intense action' in combating the spread of the disease by banning all travel from China, accompanied by his claim that 'the EU failed to take the same precautions' taps into precisely the same principles.

Fight versus Flight. Us versus Them. Right versus Wrong.

In the UK, let's revisit the Brexit debate and examine the central thrust of the argument put forward by the successful Leave campaign. Europe is bad for us. The demands that Brussels makes on us are detrimental to national security, to our economic integrity and to our sense of cultural identity. Time to make a stand.

Fight versus Flight. Us versus Them. Right versus Wrong.

And then there's the Isis terror suspect we encountered in that simulated interview and his closing words: 'Our God is holier than your God. Our warriors are mightier than your warriors. Our good is deeper than your bad.' Consider those words alongside an extract from *The Management of Savagery*, a key Isis manifesto written by the Islamist strategist Abu Bakr Naji in 2004:

> By polarization here, I mean dragging the masses into the battle such that polarization is created between all of the people. Thus, one group of them will go to the side of the people of truth, another group will go to the side of the people of falsehood, and a third group will remain neutral, awaiting the outcome of the battle in order to join the victor. We must attract the sympathy of this group and make it hope for the victory of the people of faith, especially since this group has a decisive role in the later stages of the present battle.

Or, alternatively, consider them alongside Winston Churchill's iconic wartime peroration to the UK Parliament on 4 June 1940, under threat of German invasion:

We shall go on to the end. We shall fight in France, we shall fight on the seas and oceans, we shall fight with growing confidence and growing strength in the air, we shall defend our island, whatever the cost may be. We shall fight on the beaches, we shall fight on the landing grounds, we shall fight in the fields and in the streets, we shall fight in the hills; we shall never surrender, and if, which I do not for a moment believe, this island or a large part of it were subjugated and starving, then our Empire beyond the seas, armed and guarded by the British Fleet, would carry on the struggle, until, in God's good time, the New World, with all its power and might, steps forth to the rescue and the liberation of the old.

Fight versus Flight. Us versus Them. Right versus Wrong.

Formulate a case around these three binary axes, set out your position so that it boots up and runs these three fundamental programmes of deep, ancestral categorization, and the writing is on the wall in big, bold, supersuasive letters. In evolutionary black and white. Irrespective of what you might say, and regardless of how you might say it, others are going to listen.

It's a trick that even a ten-year-old can master.

The three evolutionary ages of black-and-white thinking

One evening, not long after I'd first arrived in Oxford, I found myself at a formal college dinner sitting opposite a paleontologist friend. Just as the hors d'oeuvres arrived he asked me what I was up to, and by the time the plates were being cleared away at the end of the main course he was fully up to speed on antediluvian super-frames, ghostly categorical footsteps in lost prehistoric sands and archetypal superchords of influence. When, at last, the cheeseboard made an appearance, he made it perfectly clear what he thought about it all.

'Fascinating,' he observed. 'Truly fascinating. But are you absolutely sure that you've really discovered these super-frames? I mean, *genuinely* discovered them? Not merely, how shall I put this, *invented* them?'

His point was a good one and deserving of serious thought. It's so easy to think that you've unearthed something original when in actual fact you've buried it there yourself. Some years ago now I recall watching a television documentary about an intrepid, indefatigable ghostbuster hot on the trail of crop circles. There he is, hovering above a field in a helicopter when, sure enough, one miraculously forms beneath him on the ground . . . precipitated by the action of the rotor blades. Unable to make the connection, he can't believe his luck.

Was the same thing going on here? Was the theory driving the data? Or were the data, as it should be, driving the theory?

Fortunately, I'd put in enough research miles to be confident it was a case of the latter. The first of these super-frames – Fight versus Flight – we share with all other living species. Albeit these days, of course, with a bit of a modern makeover. Our brains process challenges to our authority, self-concept and worldview as now-or-never, all-or-nothing, life-or-death imperatives setting off evolutionary smoke alarms in higher-order cognitive centres and shutting down perspective and reasonable, rational argument. The outcome is painfully familiar. Act first. Think later. And more often than not, regret it. When the heat is on, the temperature rising, and we depend upon them most, we quite literally 'lose our minds'.

But if 'Fight versus Flight' is a super-frame common to all biological organisms, then its two more recent accompaniments are exclusive to just the one. No sooner had natural selection gone down the road of slotting big, powerful brains into our soft, slow, puny little bodies, than it needed to ensure that those bodies stuck together. 'Us versus Them' served to facilitate in-group cohesion at a time when our prehistoric ancestors first began cohabiting in small groups (*c.* 2.5–3,000,000 years ago), while as language and consciousness emerged and both group size and group number increased (*c.* 100–300,000 years ago) 'Right versus Wrong' acted as a further spur to cohesion as an instrument of social control (see Appendix V, p. 323). These simple principles of primeval categorization comprised a three-lane cognitive thoroughfare upon which our treacherous evolutionary road trip through natural selection's wild, unpredictable bandit country ever so circuitously unfolded.

Bumpy and potholed, it got us to where we are. But in recent times that road has become a lot faster and much busier. Offshoots have proliferated exponentially, and big, scary intersections have sprung up out of nowhere. The world of our forebears was a lot less complicated than the one we currently occupy. Its diameter spanned just a few dozen miles for a start. Societies were small, numbering no more than fifty to a hundred members, and were isolated from other groups. The environment was stable. Challenges were simple and short-term. And categorizations clear-cut. Antelopes were to be speared (*Fight*); lions dodged (*Flight*). Kinsfolk helped (*Us*); encroaching tribes resisted (*Them*). The spoils of the hunt shared (*Right*); the mates of other members of one's tribe avoided (*Wrong*).

Our decision-making, in other words, evolved to negotiate clear, clean, black-and-white trajectories unencumbered by context and circumstance; to cut a straight, unbroken path to safety, success and survival. And such has been the utility of this evolutionary freeway, taking us from the dawn of Cambrian prehistory to the cradle of Trumpian post-truth, that still it continues to function. Such were the benefits of these three categorical traffic lanes of ancestral cognitive necessity that still, to this day, they remain navigable and open for use. And persuaders and influencers familiar with the lie of the land take full and complete advantage.

Today, however, we live in an age when traffic in the 'Us versus Them' lane is especially congested. It has become the go-to lane for an increasingly intolerant, progressively myopic number of the world's major 'truth' stockists and wholesalers – governments, lobbyists, politically partisan celebrities – to transport their hashtags and distribute their handles and key words. Such binary bottlenecks and interminable tribal tailbacks mark a significant departure from how things used to be. Not so long ago, the conveyance of guilt and ethical responsibility was confined to the 'Right versus Wrong' lane; there, on the expressway to conscience, convoys of principled, accountable action revved and rumbled to a vanishing moral horizon. But truth, reality and objectivity have since changed course in unison like some synchronized metaphysical motorcade. These days, for instance, rather than debate the rights and wrongs of a major global incident, like the chemical attack on Syrian rebels in Douma in 2018, those sympathetic to such an atrocity contest that it ever happened, claiming, as have the Assad denialists, that it was, in fact, a fiendishly devised propaganda exercise by the Syria Civil Defence, or White Helmets* as they're more commonly referred to in the media, precision engineered to incentivize jihadists and terrorists.

Other examples aren't hard to come by. Take the website *LyinComey.com* set up by the Republican National Committee to coincide with the release of the former FBI director James Comey's memoirs. Comey, in a resplendently undemocratic manoeuvre, was relieved of his duties by Trump back in the spring of 2017 for sailing too close to an iffy, whiffy, defamatory Russian wind, and there obviously existed a modicum of bad blood between the two

---

* The White Helmets are a volunteer organization offering humanitarian aid in rebel-held areas of war-torn Syria. They undertake search-and-rescue operations in the aftermath of offensives conducted by the Assad regime and are widely credited with saving thousands of civilian lives.

gentlemen around that and other incidents. There is also the consideration that any autobiography, let alone one so clearly derived of the score-settling genre, is always going to be served up with a liberal assortment of ego-preserving hubris, self-aggrandizing flattery and mitigatory exculpations. But, even in light of the acrimonious backstory surrounding the book's release, what was still particularly noteworthy about the White House's reaction to it was just how casually on the one hand, and resolutely on the other, the *facts* of Comey's assertions were dismissed. *LyinComey.com* didn't just trash Comey. And it didn't just trash his version of events. It trashed *all* versions of events. It trashed the fact that the events had even happened.

A month or two earlier, it was possible to discern a similarly noxious ideology in the aftermath of the mass shooting at Marjory Stoneman Douglas High School in Parkland, Florida, which left seventeen dead and injured seventeen others. Wholly unsatisfied with a nod to the Second Amendment and a perfunctory, if impassioned denouncement of proposed gun-control measures, a cacophonous chorus of rightists decided to kick things up a notch by claiming that the massacre had not, actually, taken place at all. And that the grieving parents, and distraught, disconsolate teenagers, were a cast of 'crisis actors'.

Descartes was mistaken. It's not so much, 'I think, therefore I am.' More, 'We are, therefore I think.'

Perhaps unsurprisingly in an age of Twitter, Facebook, Instagram and all the other social networking platforms available to us, such a phenomenon, in which the truth or falsity of a statement is contingent not upon its verifiability but, rather, upon whether the person making it is considered one of us or one of them, has already acquired a name. 'Tribal epistemology', according to the American blogger David Roberts, refers to a biased mode of thinking in which 'information is evaluated based not on conformity to common standards of evidence or correspondence to a common understanding of the world, but on whether it supports the tribe's values and goals and is vouchsafed by tribal leaders. "Good for our side" and "true" begin to blur into one.'

But while Roberts may well be on to something, the premise isn't new. In fact, it's been with us for quite a while. Our ancient capacity for tribalism has always exerted a direct and powerful influence not just on the way we think about the world, but on how we actually *see* it.

Ever since the days of our ancient cave-dwelling ancestors, 'we' has served as the acceptable face of 'I'.

Division, decision, derision

One blustery Saturday afternoon back in November 1951, the Dartmouth Indians and Princeton Tigers – two US college football teams – squared off against each other in Princeton's Palmer Stadium. It was the last game of the season, and Princeton were undefeated. A few minutes after kick-off it quickly became clear that it wasn't going to be one for the purists. Tempers flared and fists began to fly. In the second quarter the Princeton quarterback had to leave the field with concussion and a broken nose while, in the third quarter, his opposite number sustained a broken leg.

A few days later, one of the Princeton college rags, *The Princeton Alumni Weekly*, called the game a 'disgusting exhibition' and opined that 'blame must be laid primarily on Dartmouth's doorstep'. Their team might've pulled their punches, or so the *Weekly* reported, but page and pitch were very different ball games and now that the match was mercifully over and done with, and legs and noses safely confined to plaster, the battle of the newsrooms was about to begin in earnest. The typewriting gloves were well and truly off.

Oddly enough, the Dartmouth student paper, *The Dartmouth*, saw things rather differently. Yes, the game was physical, it acknowledged. There was no denying that. But the injury to the Tigers' quarterback went with the territory. Besides, it pointed out,

Princeton's sanctimonious one-armed-waiter riff – that they took it but didn't dish it out – was at best disingenuous and, at worst, the biggest load of poppycock they'd heard all year. Or words to that effect. The Tigers, it alleged, during the course of their unbeaten season, had *themselves* taken out a number of their opposition's star players in what appeared, in hindsight, remarkably like a strategy.

Debate continued to rage and fingers continued to point long after the final whistle. But not all of them, it transpired, in anger. The sheer magnitude of the divergence between the two opposing camps not just in opinion but seemingly in basic, physiological perception – in what they actually *saw* – inspired some at the feuding colleges, the psychologists, to do something about it. To join forces, no less, to see if they could get to the bottom of *why*, exactly, *how*, exactly, the respective members of these two prestigious schools could themselves almost come to blows over the supposedly objective proceedings of an end-of-season football game.

A week after the match they administered a questionnaire to both Dartmouth and Princeton students. Who started it? Surprise surprise, most of the Princeton cohort were of the same mind – the visitors had kicked off first – whereas the Dartmouth contingent, to be fair, held both sides equally responsible. But there was a problem with this line of inquiry. Such differences in perspective might easily be attributed to nothing more revealing than the vagaries of selective memory. Or even – who'd have thought it? – exposure to biased reporting.

So the researchers did something else. To get round the issue of bias and preconception they invited into the lab two new batches of students – one from Princeton, the other from Dartmouth – and this time presented them not with a blank-slate 'witness statement' questionnaire that relied solely on subjective recall, but with a tape of the game itself.

Who started it *now*?

The response was astonishing. Even with the facts right there in front of them, even with the evidence staring them in the face, students from the two institutions continued to butt heads over what had actually happened. The authors were well and truly dumbstruck, concluding that: '. . . the "same" sensory impingements emanating from the football field, transmitted through the visual mechanism to the brain, also obviously gave rise to different experiences in different people.'

In fact, the game, they surmised, was 'actually many different games' and that 'each version of the events that transpired was just as "real" to a particular person as other versions were to other people'.

Seeing isn't believing. Seeing is *belonging.*

It isn't difficult to appreciate the power of what social psychologists call the 'in-group bias'. Show up at pretty much *any* football game – or just spend five minutes on Facebook – and you'll soon get the hang of it. Yet both its importance and pervasiveness in everyday life is often underestimated. Our affiliations with groups – from friends to family to sports teams to political parties to nationhood – have a deep and lasting influence on how we perceive, interpret and respond to the actions of those around us, often with consequences that far exceed the compass of arbitrary, one-off encounters.

To illustrate, recall a little earlier how ten-year-old Josh had informed his mother that 'Spencer's mum' – whom she'd known from her own schooldays – 'was allowing him to grow *his* hair'. And how immediately afterwards he'd added, seemingly for good measure: 'When we were in the car the other day she told his dad that making him cut his hair when he didn't want to was, like, abuse.'

What was he up to? Why did he do that? The reason, as we now know, was due to his indisputable standing as evil psychological genius, a standing, I'm sure, to which many a frazzled parent can attest. He was, quite deliberately, activating the social identity, triggering the interpersonal category, 'best friend' (Spencer's mum).

And what do best friends generally tend to do? See things in the same way. Be on the same wavelength.

The truth that Josh had instinctively come to realize was a profound one. The need to belong is stronger than the need to be right. Factually. Morally. Mentally. Physically. In every sense of the word. Reality is in the mind of the perceiver. And the mind of the perceiver is often in the mind of other like-minded perceivers.

'Us' and 'them' are about as black and white as it gets.

Several years ago a team of researchers from Yale, Temple, George Washington and Cornell Universities in the States conducted a study remarkably similar to the one performed on those hallowed collegiate playing fields of Ivy League New Jersey back in the fifties. But this time the ball was moral and ideological. And it was sober-minded Republicans and Democrats, not hot-headed Dartmouth and Princeton fans, who were sitting in front of the telly.

The footage they were shown was taken from a political protest. It was identical and standardized for both. But there was, as you'd expect, just a little bit more to it than that. Both the Republicans and Democrats were split into two camps. One camp was told that the protesters were of a liberal persuasion and were united against the ban on gay and lesbian soldiers serving in the military (such a ban existed at the time). The other was told that the activists were pro-life conservatives and were demonstrating against legalized abortion. In other words, both Republican and Democrat sympathizers were shown footage that they believed was in line with their political sensitivities and also footage that they believed was not.

The question, of course, was would any of this make any difference? Could political affiliation, like sporting allegiance and loyalty, actually influence what they *saw*? Could it make them doubt the evidence of their own eyes?

The answer was yes. A big, binary, black-and-white yes. Republicans, it transpired, looked with considerably more favour upon the prospect of police intervention when the demonstrators were ostensibly liberal and protesting against the injustice of homophobic military recruitment than when they believed them to be fellow conservatives and marching against abortion. The reverse held true for the Democrats.

Just as with the Dartmouth and Princeton fans, just as with Josh and his mum, 'we' was in the 'I' of the beholder.

Who we are determines what we see.

Persuasion mix

Back in the early seventies the late Henri Tajfel, a psychologist at the University of Bristol, performed an experiment which immediately set the tone for pretty much any work on groups conducted since. Unsurprisingly it's become a classic, lending its name to an eponymous and much used paradigm within the field of social psychology: the minimal group paradigm.

Tajfel began by taking a bunch of high-school students and showing them a display of dots. 'How many dots do you see on the screen in front of you?' he asked each of them in turn. Because there were quite a few and the time allowed was less than half a second, the students were totally clueless. They had little idea how close they were to guessing the actual number. But they provided their estimates anyway, completely unaware that the set-up was part of a brilliant and cunning plan to drive a temporary wedge between them, an ingenious device to divide them up into two arbitrary 'minimal groups': under-estimators and over-estimators. 'Minimal' because the groups had essentially been assembled out of thin air. 'Groups' because that's what they were.

With the students nominally divided and the categorizations duly made, Tajfel got down to business. Each student, he instructed, was to allocate points – which he informed them equated to money – to two of their fellow participants in the study. These other students were anonymous. They were identifiable solely by code and by one or other of the following pair of labels: 'Of your group'/'Of the other group'.

Would the simple fact that the students were members of one group as opposed to another influence their distribution of points? The answer, exactly as Tajfel had predicted from the outset, was yes. Even though the participants knew each other well – they were actually recruited from the same year at the same school – and would, after the study, return to their natural social order and interpersonal habitat, promises of financial reward were thrown around among members of one's own group like winning in Vegas while eluding those of the other group completely.

Goodies, unashamedly, were doled out in black and white. My lot as opposed to yours.

Now, if none of this comes as any great surprise to you, then it's no great surprise. It's not difficult, when you look at the world, to appreciate the hold that the need to belong and the need to fit in has over all of us. And it's not difficult to see how if the use of black-and-white words like 'brilliant' and 'horrific' can nudge us, as we saw in Chapter 8, into being more extreme, then the cognitive activation of black-and-white super-categories like 'us' and 'them' has the capacity to do the same. But what *may* raise an eyebrow is the sheer extent to which presenting an issue from an Us–Them perspective, within an Us–Them *frame*, can bring people around to our own way of thinking. And, moreover, can harden their resolve to act on their newfound conviction.

To illustrate further, consider a fairly recent study conducted by social psychologist Nik Steffens of the University of Queensland. Steffens scrutinized the content of the election speeches made by leaders of the main Australian political parties dating back to Federation in 1901 and discerned a secret message concealed within the language. Leaders who eventually went on to win those elections used the pronouns 'we' and 'us' on average once every seventy-nine words, whereas the losers used those same terms only once in every 136 words. In addition, in thirty-four of the forty-three elections he analysed it was the candidate who invoked the idea of 'we' and 'us' most frequently who ultimately ran out as victor.*

Whisper it quietly in the corridors of power. Talk not of what *you* think but rather about what *we* think.

Let's look at the flip side: at the kind of thing that happens when one disrupts the 'us/we' dynamic within a team and alters the psychological stage lighting so that regular members of the cast are consigned en masse to the shadows while one or two high-profile

---

* Research into the link between pronoun use and personality style indicates that leaders' greater inclusion of the word 'we' in their speeches might not be the only contributing factor behind their comparative election success. Associated character traits that correlate highly with increased use of the 'we' pronoun may also have something to do with it. Studies demonstrate, for example, that extroverts use words such as 'we', 'our' and 'us' more than introverts who, in contrast, tend to favour the more singular and disjunctive 'I', 'me' and 'mine'. Further evidence suggests that an increased frequency of the word 'I' within certain contexts alludes to heightened feelings of threat, insecurity and defensiveness – qualities not typically associated with charismatic personalities and inspirational leadership. In 2007, the American psychologist James Pennebaker – in partnership with the FBI – scrutinized the post-9/11 correspondence of al-Qaeda leader Osama bin Laden and the post-2003 invasion of Iraq correspondence of his second in command, Ayman al-Zawahiri, while both men endeavoured to evade capture. Results revealed a striking disparity between the communication patterns exhibited by the two jihadi figureheads. Firstly, bin Laden's use of so-called 'exclusive' or qualifying words such as 'except', 'but', 'however' and 'without' – words generally associated with reduced black-and-white thinking and enhanced cognitive complexity – increased over the course of the fifty-eight transcripts analysed, while that of his more 'binary' second officer decreased, a tendency possibly indicative of the latter's elevated need for strict ideological closure under stress. Second, while al-Zawahiri's use of the 'I' pronoun escalated threefold in the wake of the Iraq War bin Laden's remained unchanged, a trend which offers a unique insight into the al-Qaeda founder's more visionary style of leadership – his deeper appreciation of the bigger picture – as well, quite possibly, as his greater presence of mind. (Note: in May 2011, following the death of bin Laden, al-Zawahiri became the new leader of al-Qaeda, a position he still holds today.)

individuals bask in the value spotlight. If you're the manager of a large, successful football club with a big fat cheque book burning a hole in your pocket this is precisely the type of dilemma you face every season during those critical, uncertain times when the transfer window opens and you have leave to draft in new players. The way things stand at the moment, the go-to strategy in most cultures and organizations seeking to raise their game and increase their competitive advantage – be they in sport, in business, even in academia – is to buy in talent from outside. But be warned. There are problems with big-name signings. And some pretty good science to prove it.

Some twenty years ago now, organizational psychologist Matt Bloom from Notre Dame University in the States conducted a study of twenty-nine major league baseball teams in Canada and the US. What he discovered should send shivers down the spine not just of football and baseball-loving oligarchs the world over but of bosses and team leaders everywhere who are trying to get the best out of the groups and individuals they work with: a marked correlation, over the course of an eight-year period, between high levels of intra-group disparity in pay (usually precipitated by the arrival of a few highly paid stars) and a significant decline in both individual and team performance. It's a simple lesson to teach but a difficult one to learn. *Us versus Them* may well get things going *between* groups – as Isis and Churchill attest. *Within* them, however, it's deadly.

But the appliance of our three evolutionary super-categories – Fight/Flight, Us/Them, Right/Wrong – to the bending of wills and the changing of hearts and minds isn't just exclusive to the nurture of *optimal* performance. While they're certainly effective at starting fires, they can also be used to extinguish them. No matter how hard you work or how much you try – as an individual, as a team or as part of a large organization – there will always be problems at some point, and when relationships break down and difficulties and differences arise, they may also be deployed in the reduction and resolution of conflict.

Nowhere is this better illustrated than in a recent episode involving Sean Dyche who, you'll recall from Chapter 5, is currently in the hot seat at Premier League football club Burnley. In September 2019, just a handful of games into the new season, Danny Drinkwater, the England midfielder who was playing for Burnley at the time, got into a spot of bother. Physically set upon by half a dozen men outside a nightclub, he ended up injuring his ankle. Out of action for a month.

Drinkwater wasn't a permanent fixture at Burnley. He was on loan to the club from Chelsea. He'd also been convicted, just a few months previously, of drink-driving. Dyche, had he wished, could easily have been forgiven for reading the player the riot act. But he didn't. Instead, when quizzed by the media as to whether he'd taken any action against Drinkwater, the manager decided to stand by his player, pledging support for the troubled midfielder and promising to take whatever steps were necessary to get his career back on track.

Here's a transcript of what Sean had to say about the matter at the time:

> First, I wanted to find out the facts, which I did. Secondly, look at the reality of it and forget about him being a footballer.
>
> They are human beings and have private lives which are sometimes not private. Sometimes people get in scrapes.
>
> I think he knows that at his age – he's twenty-nine – he has to mind what he's doing and be in the right place rather than the wrong place.
>
> My responsibility is to give him the chance to take action and lay down some of the guidelines, his responsibility is to deliver it.
>
> He's had a tough couple of years football wise, not playing much. It's as much down to him and his drive and his

will and desire – not about the incident, but to get back to being a top footballer. He wants the hunger and desire to flow back into him, but he's got to earn that. It's not a given.

Sometimes when things aren't right for you on the pitch they do manifest themselves into something off it too. It can drop you out of kilter a little bit.

So I'll be working with him, getting him fit, helping him, guiding him back to where he wants to be.

It's easy to spend time with players when things are rosy, sometimes it's more satisfying to work with players when things aren't right.

Every manager wants players with a great attitude who are flawless and they all come in bright and breezy, play football and then go home and eat chicken and pasta and drink loads of water.

It's just not the case. They are human beings. I've got flaws, they've got flaws.

Pretty much everything Sean articulates provides our brains with a warm, buzzy, 'instant hit' black-and-white-thinking fix of *Fight versus Flight, Us versus Them, Right versus Wrong* cognition. Not, perhaps, in the way we've grown accustomed to up until now: the hot, hard, badass evolutionary stimulant of a radical call to arms, a persuasion and influence 'upper'. But in compassionate, conciliatory form. Deftly, insightfully, responsibly, Sean's empathetic, magnanimous response, on closer psychological inspection, hits all three dials on our supersuasion 'mixing desk'.

Nine times out of ten in a situation like this the natural thing to do is to lash out. To throw the book at the player. Fine him. Drop him. Or get rid of him altogether. But is it the *right* thing to do? The smart thing to do? Taking emotion out of the equation, what's the best course of action for the individual concerned? Or, for that matter, the club? Turn your back on them and say: 'You're on your own'? Shunt the 'them' control up to max, in other words. Or do the opposite? Flick the 'us' switch into overdrive and stand shoulder to shoulder with them as they confront their errant behaviour, examine their wayward actions and work through their underlying issues?

Sean, as we've seen, goes for the latter option. He dials up the 'fight' button not against the player but against the demons they're doing battle with. And he chooses his words carefully. On two separate occasions he categorizes Drinkwater not as a 'footballer' but as a 'human being', removing, at a stroke, not just any *moral* barrier that may be perceived to exist between them, but also, for good measure, the barrier between player and manager: 'I've got flaws, they've got flaws.'

Whatever your views on Premier League footballers the message is loud and clear. Whether you're trying to talk someone *off* a ledge, as our teachers did with the Taliban in Chapter 8, or whether you're trying to talk someone *on to* one, as Churchill did in Parliament during the war, your immediate success as a persuasion and influence 'producer' will inevitably depend on your technical expertise on the three-track supersuasion mixing desk.

Cast your mind back to the scene that greeted that Taliban commander and his men when they gate-crashed that classroom in the school in Pakistan. The welcoming tea and cakes reception instinctively dialled up 'us'. The subordinate 'better wives' riff immediately dialled up 'right' (it was Peshawar, remember, not Peterborough). And the 'better understanding of the Qu'ran' routine instantly dialled up 'fight'.

The teachers got exactly what they wanted for one simple reason. Touched by the hand of lost primordial genius they hit all three buttons on the supersuasion mixing desk. *Fight versus Flight. Us versus Them. Right versus Wrong.*

The three ancestral hallmarks of every silver tongue.

CHAPTER 11

Undercover Influence: The Secret Science of Getting What You Want

∴

The secret of my influence has always been that it remained secret.
SALVADOR DALÍ

At daybreak the morning after we had repelled an attack, we saw that the Germans had collected their wounded, with the exception of one man who lay groaning in agony halfway between the trenches.

Our captain jumped forward from his trench. The Germans fired and he was hit.

He staggered but with a magnificent effort kept his feet and rushed on towards the wounded German. Then, although badly wounded, he picked him up and carried him direct to the German trenches.

We heard the roar of cheers as he gently laid the body down and saluted. A German officer climbed up from his trench and, removing his own Iron Cross, pinned it on our hero.

Sadly, back in our lines the captain died from his wounds

and I am broken-hearted that his cross is a wooden one rather than the VC he deserved.

From a letter published in the *Daily Mail*, 11 November 1914.

Last summer, on a weekend away down in Cornwall, my wife and I were out for a country stroll. Breezy sunshine filtered through the trees, the cows were snuffling behind the hedges, and the lazy, hazy air smelt like the very embodiment of wholesome rural calm. I half expected John Craven from *Countryfile* to suddenly pop up from behind a bush brandishing a punnet of strawberries and a sparkling flute of something made from elderflowers.

Instead I get a cross between *Dad's Army*'s iconic windbag Captain Mainwaring and Joaquin Phoenix's embodiment of the Joker.

'Oi! You! This is a private road!'

We turn to see a man stomping down the lane towards us. He doesn't have a blunderbuss. Nor does he say, 'Get orff my land.' But with the brogues, plus-fours and hunting jacket he's the picture of a country squire.

'Sorry, we didn't realize it was private,' I explain. 'Look, we're almost at the end of the road now. So how about we just scoot along and we'll be out of your way?'

This eminently reasonable suggestion elicits a short, contemptuous laugh.

'Oh, I seeee . . . so you reckon you can just do that, do you? Just continue down the road on your merry little way? Well, here's a thing: you can't! I want you off my property. Now! Turn around and go back the way you came.'

My wife and I exchange glances. Surely this doesn't make sense. If the deal was simply to minimize wear and tear on his tarmac then the sensible option would be to let us carry on. But we shrug, turn on our heels and retrace our steps until we're back to where we started. Three-quarters of a mile down the other end of the road.

If you needed any convincing that black-and-white thinking isn't black and white but is itself on a spectrum along which each of us has our place, then these two accounts should be right up *your* street. When that wounded British captain braved the mud and the blood of no-man's-land to rescue the German soldier, he didn't think twice about the line he was going to cross. In an act of extraordinary selflessness, he completely disregarded the labels 'British' and 'German' and thought only in terms of the broader category, 'soldier'. The farmer, on the other hand, had parcelled up his world into the categories 'mine' and 'yours' with the slavering, blinkered tenacity of one of his sheepdogs latched on to a bone.

It may sound as if I'm judging these men. But I'm not. On a different day that gallant British captain might've looked after number one. The farmer, if he'd been in a sunnier mood, could well have thought, 'Who cares? Let them get on with it.' We'll never know. But what we *do* know is that sometimes we have tunnel vision. And other times we don't. Sometimes we think like torches in wide arcs of open-minded reason. Sometimes we think like lasers in narrow, fundamentalist beams. It all comes down to our viewfinder. Where it's pointed and how it's set. The captain's was adjusted to close-up, blurring the edges of the us-them, me-you divide. The farmer's was adjusted to landscape where such boundaries are clearly defined.

We have the Greek poet Archilochus to thank for first conceiving of the possibility that the chequerboard mindset of black-and-white thinking might fit along a spectrum. Employing a metaphor that anticipated the notion of the monochrome mind by some two and a half thousand years, he divided people, in decidedly binary fashion, into two opposing kinds. Foxes and hedgehogs.*

---

* See Appendix VI, p. 327, for cultural and historical differences in black-and-white thinking though the centuries.

'The fox knows many things,' declared Archilochus. 'But the hedgehog knows one big thing.'

Much later, in the mid-1900s, the Latvian-born philosopher Isaiah Berlin expounded on the analogy. The strength of the hedgehog, observed Berlin – Plato, Dante, Hegel and Nietzsche included – lies in his focus and constancy of vision. His passage through life is overwhelmingly determined by slavish allegiance to just the one, overarching principle. The power of the fox, on the other hand – Aristotle, Shakespeare, Goethe and James Joyce, among others – lies in his flexibility and openness to experience. Unlike the hedgehog, the foundations for his actions are manifold, Berlin proposed. They cannot be reduced to just one, all-encompassing worldview. The hedgehog never wavers. Never doubts. The fox is more expedient and pragmatic. And more inclined to see complexity and nuance.*

Research into the field of psychological disorders tends to corroborate Archilochus's musings and suggests that we're talking about a spectrum of black-and-white thinking; that the all-or-nothing continuum of cognition might be anchored at either end by a pair of dysfunctional mindsets as different from each other as black and white themselves.

At one extreme of this clinically dichotomous dimension, this sliding scale of overly fluid versus excessively bounded self, we have psychosis, where all things are connected, all things are personally relevant, and all things are imbued with deeper significance and meaning, manifesting as delusions, hallucinations and disorganized speech and behaviour. At the other we have either/or and the

---

* In his international bestseller, *Thinking, Fast and Slow*, the Nobel laureate Daniel Kahneman argues a similar case for there being a dichotomy between, as he calls it, Type 1 and Type 2 cognition. The former is characterized by fast, instinctive, intuitive and emotional thought processes, while the latter variety is slower, logical and more analytical in nature.

world of the brutally disjunctive: enslavement to order and rigid, rule-based behaviour, the far reaches of which might naturally encompass autistic spectrum disorders.'*

Most of us, of course, fall somewhere in the middle. Our abstract and analytical minds room together quite happily in the buzzing cognitive dormitories of our brains. But what underlies such natural variation? What are the secret arbiters of grey? What is the mechanism that drives this dysfunctional duality, the monochrome ghost in Archilochus's fox–hedgehog machine? The answer, it turns out, takes us back to familiar territory, to differing levels in our need for *cognitive complexity* that we learned about from Arie Kruglanski in Chapter 6. Let's revisit this notion again. But this time in a little more detail.

Cognitive complexity is defined in terms of two variables: *differentiation* and *integration*. Complexity of differentiation refers to the

---

* Black-and-white or 'all or nothing' thinking is an underlying feature of a number of mental health issues, not just those experienced by sufferers of autism or autistic spectrum disorders. Perhaps the best documented of these is the so-called 'catastrophic cognition' that one observes in many recovering addicts in the face of a minor relapse. Just the one drink, a single fix or a couple of drags on a cigarette are often construed as completely undermining weeks, months and sometimes even years of unblemished, hard-fought abstinence and frequently heralds the return of the unwanted habit in all its former guilt-inducing glory. To combat such all-or-nothing thinking, counsellors will typically endeavour to interpose shades of grey between the black and the white pillars of the addict's extremist mindset, sometimes resorting to metaphor. Analogies such as 'everyone falls off at least once when learning to ride a bicycle' and 'it's just one stop on the line – you don't have to go all the way to the end' prove effective in this regard. On the flip side, if you're addicted to tough decision-making, then catastrophic, black-and-white thinking can actually be more of a blessing than a curse. In his bestselling book *Leading*, Sir Alex Ferguson, the legendary former manager of Manchester United, tells the story of how, on some nights, when he was the manager of Aberdeen, he and his assistant Archie Knox would often face a round trip of six hours to watch opposition games in Glasgow, sharing the driving between them there and back. 'Whenever we got tempted to skip a game and take the night off,' Sir Alex recounts, 'we'd always say to each other, "If we miss one game in Glasgow, we'll miss two."'

number of characteristics or dimensions of a problem that are taken into account when considering an issue. When an individual thinks in good/bad terms, for instance, in binary alternatives, he or she is clearly thinking in an undifferentiated manner. High differentiation, in contrast, occurs when a person views an issue from multiple perspectives and with a fair degree of nuance.

The complexity of integration, on the other hand, depends on whether the individual perceives such differentiated characteristics as operating in isolation from each other (low integration) or in multiple contingent patterns (high integration). Thus a decision-maker with a high need for cognitive complexity will typically evaluate all the relevant perspectives on an issue and then integrate them into a coherent position, while one with a low need for complexity will generally entertain only the one viewpoint and maintain it with dogmatic tenacity.

Take three children each presented with a helping of Brussels sprouts. The first pushes their plate away with a scowl of disgust: 'I hate Brussels sprouts.' That child has one simple, undifferentiated perspective on Brussels sprouts. The second prods dubiously at the plateful: 'I like how it tastes but the texture is horrible.' That child entertains a more complex and differentiated view of the vegetable, holding two distinct ideas about it at once. The third child takes a more nuanced approach: 'Brussels sprouts have a nice flavour and a horrible texture. But it's the way the two combine in my mouth that makes the experience of eating them so unique.' That child not only harbours two ideas about Brussels sprouts at once, but also introduces them to one another to create a third. That's integration – and the winner of *Junior Masterchef* – in action.

Why we need extremists

There is evidence to suggest that differences in the need for cognitive complexity are found across all sorts of divides, not just when it comes to the symptoms of mental disorder (or indeed, vegetable preferences).

In politics, for instance, research has shown that Republicans, on average, present with a lower need for cognitive complexity than Democrats. The liberal-conservative fault line is a cognitive complexity frontier. Those who have a high need for structure, and are less inclined to be open to new experiences, tend to be small 'c' conservative in outlook. That's why, if you're a big 'C' Conservative, you're more likely to trust in tradition and want to preserve the status quo. Change equals uncertainty and a blurring of the lines. You're also more likely to think in absolute terms about all kinds of topics than a Liberal with a capital 'L' (or, for that matter, one with a small 'l'), whose tolerance of ambiguity and complexity will in all likelihood be greater.*

Religion, unsurprisingly, is another case in point. Extremists, as you'd expect, exhibit a lower need for cognitive complexity than moderates. Radicals, studies have shown, often see the world as divided into binary groups, a good example being the Protestant fundamentalist culture of distinguishing people according to those who are 'saved' and 'unsaved', 'sheep' and 'goats', 'lost' and 'found'.

---

* In addition to differences in the need for cognitive complexity, it would also appear that conservatives and liberals vary in terms of certain elements of their core personality structure as well. Research shows that among the so-called 'big five' personality variables (openness to experience, conscientiousness, extroversion, agreeableness and neuroticism) the first two significantly correlate with ideological tendencies. Conscientiousness – which breaks down into two subcomponents: 'orderliness' (the need to keep things organized and tidy) and 'industriousness' (which relates to productivity and work ethic) – has been found to lean right while openness to experience, which involves 'active imagination, aesthetic sensitivity, attentiveness to inner feelings, preference for variety, and intellectual curiosity' tends, as intimated above, to lean left. Agreeableness, too, has also been shown to correlate with political preference but only when deconstructed into its two active ingredients: 'compassion', which is positively associated with liberalism, and 'politeness', which is positively associated with conservatism.

It's no coincidence, as we saw earlier, that the flag of Isis bears just the two colours: black and white.

But context is also important. Back to politics, and research shows that at certain times and in certain situations, black-and-white-thinking leaders tend to be the ones that we vote for: we prefer their style of either/or decision-making over more considered, analytical approaches. Such times and situations are invariably those of uncertainty. The Brexit vote in the UK, for instance. Trump in the US election. Churchill during the Second World War.* These kinds of leaders under these kinds of circumstances – where there is doubt, division and disquiet – are often more popular than other candidates, even when the latter are respected and admired and have a ground-swell of support that would ordinarily, under a different set of societal or political conditions, sweep them confidently and convincingly to power. And the reason for that is simple. People want people who *do stuff*. Or at least who are perceived to do stuff. We want presidents, prime ministers and chancellors who, above and beyond everything else, can reduce our levels of uncertainty. Who are what we call 'task-oriented'. Leaders who are positive, self-assured and aggressively, compulsively, decisive.

The irony, of course, is that black-and-white-thinking leaders are sometimes the ones who create the most uncertainty in the first place. It tends to play into their hands. Take Nigel Farage, for example, EU Referendum Leave campaigner and former leader of

---

* Shortly after Churchill replaced Neville Chamberlain as British Prime Minister in May 1940, the American writer and publisher Ralph Ingersoll reported later that year: 'Everywhere I went in London people admired [Churchill's] energy, his courage, his singleness of purpose. People said they didn't know what Britain would do without him. He was obviously respected. But no one felt he would be Prime Minister after the war. He was simply the right man in the right job at the right time. The time being the time of a desperate war with Britain's enemies.' Sure enough, a mere seventy-nine days after VE Day, Labour's Clement Attlee swept Churchill aside in the 1945 general election to assume power in a landslide victory. (Ingersoll, p. 127.)

the UK's Brexit Party. Farage distils politics into its most basic ideological principles, factorizing issues into their lowest common
denominators. This may work in the short term. But, as we saw
with Brexit, one of the 'easiest [deals] in human history' as the government's then Secretary of State for International Trade Liam Fox
referred to it back in 2018, the situation became so utterly convo-
luted and byzantine that it took the combined and concerted efforts
of seasoned political commentators from right across the board to
explain it to an exasperated electorate.

Nigel's viewfinder is firmly set to landscape. Not always, as we
learned from Alastair Campbell, a good thing in politics. Nor, as
we learned from the heroic British soldier, in everyday life. In fact,
Nigel's viewfinder is set so far into the camera it occasionally takes
pictures of the photographic mechanism itself! Which for his polemical political narrative serves a purpose. You're either in or out of
Europe. End of story. But when that viewfinder lens extends out-
wards from landscape to portrait, when our focus zooms in from
ultra-wide angle to extreme, intense close-up, life has the persistent
and infuriating habit of becoming just that little more complicated.

Sometimes, needless to say, such complexity can be a good
thing, as with Brexit. But sometimes, arguably, as with an issue like
climate change, those very same nuances can stand in the way of
progress. The teenage environmentalist Greta Thunberg is a case
in point. Thunberg broadcasts her own special brand of black-and-
white thinking around the world, and regards having Asperger's
Syndrome as a positive in her life: her 'superpower'. The disorder
'makes me different,' she observed to BBC Radio 4's Today programme in 2019, 'and being different is a gift . . . I don't easily fall
for lies, I can see through things.'* Thunberg doesn't care about
what her many detractors think because, as she's been at pains to

---

* To listen to the interview in full visit: https://www.bbc.co.uk/programmes/
p07770t8.

point out on occasion, if they don't see the environmental catastrophe we're heading towards, then what's the point of listening to them?

Like Farage without the bombast and the tweed, her black-and-white brain with its 'intense preoccupation with a narrow subject' and its partisan 'one-sided verbosity'* is uniquely equipped to drive home her vital message.

Of course, the pertinent question in all of this is whether black-and-white thinking can *benefit* us in everyday life? Absolutely it can. It would be pretty black and white to claim it didn't. It's important to remember that natural selection doubt-proofed our brains for a reason, and such prehistoric imperatives of quick and dirty survival – *Fight versus Flight*; *Us versus Them*; *Right versus Wrong* – aren't without their uses in modern-day society, too.

During the Coronavirus crisis of 2020 Scotland's chief medical officer, Dr Catherine Calderwood, failed to comply with her own social isolation guidelines and drove forty miles on consecutive weekends from her primary residence in Edinburgh to her holiday home in Fife. Outed by the *Sun* newspaper, she resigned. Commenting on her actions in a television interview the next day Scotland's First Minister Nicola Sturgeon declared, with obvious irritation, that it wasn't the case that there was 'one rule for her and another for everyone else'. It was one of those few and rare occasions in politics when everyone agreed† – in complete contrast, a

---

* For more on the distinctive characteristics of Asperger's Syndrome, see: http://www.autism-help.org/aspergers-characteristics-signs.htm

† It's widely accepted by those on all sides of the party political spectrum in the UK that the government's 'Stay home, protect the NHS, save lives' slogan constituted one of the most effective communications in modern political history. That it uniquely enshrines, and directly reflects, the three primeval axes of supersuasion – *Fight versus Flight*; *Us versus Them*; *Right versus Wrong* – in precisely that order of presentation is no coincidence. Indeed, so effective did the

couple of months later, to Boris Johnson's defence of Dominic Cummings's 250-mile excursion to his parents' farm in Durham to get help with childcare while his wife was exhibiting symptoms of Coronavirus . . . when everyone disagreed.

Closer to home, or not as the case may be, imagine that you've got a couple of young kids and live near a busy intersection. Most parents would concede that the black-and-white dictum 'no playing in the street' is absolutely non-negotiable under such circumstances. No ifs and buts. No middle ground.

Imagine you've got an important project that you have to complete. What happens when the deadline comes around? Well, you've either finished it or you haven't, right? Shades of grey won't do you any favours. And neither will excuses. At the end of the day you're left with a stark dichotomy. Project or no project.

What about domestic violence? Justified under some circumstances, perhaps? Absolutely not, would be the answer of any reasonable person.

And, in high-pressure jobs, such as medicine and the military, the ability to make one's mind up when one's back is against the wall nearly always goes with the territory.

Stephen Westaby is one of the world's great heart surgeons. And also one of its toughest. Recently retired, he headed up the cardiothoracic unit at the John Radcliffe Hospital in Oxford for the best part of thirty years, and took on operations that would have had other surgeons pissing their pants. Such was his dedication to the cause that Westaby pissed in his *boots*, via a catheter, to maximize

---

message prove that when, in an address to the nation on the evening of Sunday, 10 May 2020, the Prime Minister Boris Johnson inaugurated the first baby steps of coming out of lockdown with the revised slogan 'Stay alert' there was widespread condemnation of the directive as being, in comparison, too vague, too general and too open to interpretation. 'Stay home, protect the NHS, save lives' is about as black and white as it gets. 'Stay alert', in contrast, allows for nuance and shades of grey.

time at the table. He garnered a reputation, in less staid, bureau-cratic times, as a swashbuckling braggadocio, wielding scalpel and saw in his rugby kit, blasting out Pink Floyd. A diagnosed psycho-path, he cruised darkened hospital corridors in the wee small hours like some kind of ruthless, predatory *anti*-serial killer, stalking the Grim Reaper to within an inch of his life, spoiling for fights and cooking up reasons and pretexts as he went. If he was fortunate enough to find any, he usually emerged victorious.

In the autumn of 2019 I interviewed Westaby in front of a packed house at the Cheltenham Literary Festival alongside fellow heart sur-geon Samer Nashef of the Royal Papworth Hospital in Cambridge. Beforehand, over coffee, Westaby told me about the difference between the great and the not-so-great in his line of work.

'The people who go on to distinguish themselves in this profes-sion,' he observed, 'are those who have the courage of their convictions and back themselves in the heat of battle. The luxury of deliberation is a commodity that surgeons, especially heart sur-geons, can often ill afford.

'Decisions are wonderful things to make when they come neatly packaged in time. But when they don't, you still have to make them. That's where the great surgeons earn their reputations. It's not through the tailoring . . . the cutting, the grafting, the stitching. It's through the decision-making. It's through the ability to put one thought in front of the other without thinking.'*

A Navy Seal colonel I spoke to a few years ago when I was writ-ing *The Wisdom of Psychopaths* made a similar observation about *his* branch of surgery. 'Think twice about pulling the trigger,' he told me without a flicker of emotion, 'and the next thing going through your head could well come out of the end of an AK-47.'

But the benefits of the black-and-white mindset aren't just

---

* Nashef's and Westaby's books both make interesting reading on the subject of decision-making in surgery. See the References for details.

confined to the brutal psychological lottery of *split-second* decision-making, to spur-of-the-moment, blink-of-an-eye cognition. They also come in useful when the game is long and hard. When time is turning a profit instead of running at a loss. When the battle becomes a war.

Sir Ranulph Fiennes, considered by many the world's greatest living explorer, and a special forces man himself many years ago, espouses the benefits of black-and-white thinking as an entry-level mindset for do-or-die adventurers. It's not a career for the weak or faint of heart. And neither for the grey of mind.

'Whenever I'm choosing people to take part in an expedition,' he tells me from his nerve centre deep in the Somerset countryside, 'I always go for character and motivation first and skills second. Skills you can teach. Character you can't. Not in the short term anyway.'

He looks, he continues, for the ability to see it through, 'to get it done, to come out the other side come hell or high water. Often both. The ability to put everything out of your mind – pain, hunger, cold, fear, fatigue – and focus on the goal you have set yourself. Being the first to plant the flag. The first to make the crossing. The first to do battle with the demons of the unknown.

'This may not accord too well with the spirit of the age, with health and safety and all the bleating about work–life balance and what have you. But consider this. You've got two climbers competing with each other to get to the top of a mountain. One keeps stopping to admire the view. The other keeps going, pushing on through the pain barrier until finally they reach the summit because that's all that matters. That's all that's on their mind. Getting to the summit.

'The former might *feel* better and will savour the vista for longer when they get there. And they might also have a better life expectancy! But the latter will have won the race. And if they suffer in the process, are battered and bruised by the elements, and exhaustion, and anything else the mountain might happen to throw at them, then, to them, to their mind, to their black-and-white way of thinking, it will have been worth it.'*

Ex-England rugby captain Lawrence Dallaglio agrees. Lawrence was part of the World Cup-winning side of 2003 that beat Australia *in* Australia with the last kick of the game. One spring morning over muffins and madeleines in a hotel lounge on the banks of the River Thames in West London, he tells me a story which pretty much sums up the power of black-and-white thinking.

'After we'd won the game, Prince Harry comes into the dressing room to congratulate us. I'm peeling off my shirt and underneath there's these big red welt marks from where I'd been given a good shoeing in the rucks by the Aussie pack. Harry notices the mess.

' "Whoa, Lawrence," he says. Or words to that effect. "Look at your back!"

'Obviously, I know what he's referring to. But to us it's just normal. In those days it was just another day at the office. You've got all sorts happening to you in a rugby world cup. For eighty minutes you're out there in the eye of a physical and psychological hurricane. The only passage through it relies on complete tunnel vision. Total self-belief and the belief in those around you.

'This is your moment. Your chance. This is what you've been working for. This is what all the blood, sweat and tears have been about for the past four years. Sometimes, even longer. There's only one winner. Only one trophy. Only one way of looking at it. Don't let them take it away from you.'

---

* Sir Ran stresses, however, that undue risks should never be taken on expeditions unless as a last resort. 'The key to success on the challenges I've undertaken,' he tells me, 'can nearly always be traced back to meticulous planning and obsessive attention to detail.'

Persuasion to go

Most of us, of course, have a little more psychological wiggle room in which to make up our minds than special forces commandos, cardiothoracic surgeons, extreme adventurers and gold-medal or world-cup-winning sportspeople. Yet still, in everyday life, there are decisions to be made and lines to be drawn. Any parent who *sometimes* lets their kids play near a busy road junction is asking for trouble. Any spouse who thinks domestic abuse is *sometimes* a good idea is clearly not relationship material. Indeed, our need to draw lines to create order, to partition reality into discrete, discontinuous categories and to make sense of the world around us, is the very thing that makes that world go round. It codes not just for our accomplishments as influencers. But our amenableness as influencees.

But what does this mean in *practice*? Does it necessarily follow that the better we are at recognizing the categorical bases of arguments, at factorizing them, as it were, into their lowest common (and most influential) categorical denominators, at accurately perceiving where lines are being drawn by others and where, in future, to draw those lines ourselves, the better we will be when it comes to changing minds and the more discerning we will be when it comes to resisting persuasion?

Without a doubt, the answer is yes. Natural selection never went in for instruction booklets. But if we came in a box with a manual, DVD and licence agreement, our proclivity to categorize and our default for black-and-white thinking would be there under 'Getting Started'. Learn how things work and they generally become easier to handle. We humans are no exception.

It all comes down to lines. To sides and fences. Perimeters and parameters. To tipping points and flipping points. We live in a disjointed, chequerboard reality in which we dissect, decouple and compartmentalize what's 'out there' – the 'hurrying of material, endlessly, meaninglessly' as the mathematician and philosopher

Alfred North Whitehead once put it – to avoid one long, buzzy, blurry whoosh of 'stuff'. Our brains break it up in exactly the same way as we chop up food for kids: we can't swallow reality whole. But if we can frame our persuasive messages in ways that mirror those three binary super-categories that we learned about in the previous chapter – *Fight versus Flight*; *Us versus Them*; *Right versus Wrong* – we're on to a sure-fire winner every time.

Take, as an example, my favourite subject: the EU Referendum Leave campaign.

It has all three.

Fight versus Flight: Let's stand up to the bonkers Brussels' edicts that they're systemically forcing upon us.

Us versus Them: We shall fight them on the beaches (over fishing quotas); we shall fight them on the landing grounds (over airport charges) . . .

Right versus Wrong: We should be in control of our own destiny, not subject to the laws of a faceless foreign entity.

And that was why Leave won the referendum. Supersuasion! It's as plain and simple as that. Whether Leave was right or whether Leave was wrong is irrelevant. Unlike Remain, its knockdown persuasive insight was to bring these three evolutionary super-categories on to their side; to turn an inestimably complex argument into a zero-resistance super-conductor that fizzed around the brains of an unsuspecting populace faster than you can say 'snap election'.

From knockdown to lockdown, let's deconstruct British prime minister Boris Johnson's address to the nation on the evening of 23 March 2020, in which he announced the further tightening of restrictions in the fight against Coronavirus.

Fight versus Flight: 'In this fight we can be in no doubt that each and every one of us is directly enlisted.'*

---

* We've seen previously how the use of tailored language can help 'frame' persuasive communications and render them more effective. Note, in this regard,

Us versus Them: 'The people of this country will rise to that challenge. And we will come through it stronger than ever. We will beat Coronavirus and we will beat it together.'

Right versus Wrong: 'To put it simply, if too many people become seriously unwell at one time, the NHS will be unable to handle it – meaning more people are likely to die, not just from Coronavirus but from other illnesses as well. So it's vital to slow the spread of the disease.'

But that, of course, is politics. Full of supersuasion black belts and arch-wizards of influence. As for the rest of us: what can *we* aspire to? Can all of us scale these dizzy persuasion heights – or plumb their depths, depending on how you look at it – during the course of our everyday lives? Can we all become supersuaders, channelling our ideas through others' neural pathways with frictionless ease and lightning, atomic speed?

Johnson's use of the word 'enlisted', a term strongly associated with military service. Indeed, media coverage of efforts to 'combat' the spread of COVID-19 are replete with military-style language with phrases like 'curfew', 'rations', 'front line', 'win the fight', 'beat the enemy' and 'wartime government' all featuring prominently. Nowhere was this more eloquently and poignantly in evidence than in the Queen's address to the nation at 8 p.m. on the evening of Sunday 5 April 2020, as the death toll from the virus continued to mount and rates of infection soared. Drawing parallels between the 'painful sense of separation from their loved ones' that social distancing was causing for people and the experiences of child evacuees during the Second World War, the ninety-three-year-old monarch offered the following message of hope: 'We should take comfort that while we may have more still to endure, better days will return. We will be with our friends again. We will be with our families again. We will meet again.' Indeed, building on the parallels drawn by the Queen between the COVID-19 pandemic and wartime experience, Professor Neil Greenberg, a senior health service adviser and world-leading expert in trauma at King's College London, subsequently pointed out that if NHS staff were not in receipt of adequate psychological support as the coronavirus outbreak subsided then they would be at similar risk of high rates of post-traumatic stress disorder as that of serving military personnel returning from a conflict.

The answer, fortunately, is yes. But just as in learning any new skill it takes practice, awareness and a little bit of time and effort.

Here, then, are a few simple tips to get you started.

Step 1. Uncover the hidden Sorites paradox that might underpin a particular argument.

To illustrate this first principle of supersuasion, consider the case of British-Bangladeshi footballer Hamza Choudhury. A talented youngster, Choudhury was only ten years old when he first experienced racial abuse from the pitch sidelines, yet he's gone on to become one of the few professional footballers of Asian descent to play in the Premier League. But then, not long after he turned twenty-one, a distinctly racist skeleton came tumbling out of his own somewhat rickety closet. A crass, inadvisable joke pertaining to black men and physical prowess that he'd tweeted when he was fifteen, and for which he later apologized, resurfaced. He was charged by the Football Association with 'aggravated misconduct', fined £5,000 and ordered to attend an education course.

The question is: should he have been punished?

Let's twiddle the viewfinder in a couple of notches and examine the facts. He was a child at that time, not even out of school, and had no idea of the responsibility he would later have as a high-profile footballer. What if he'd written that tweet when he was twelve? Or seven? Or five?

Where do we draw the line? What defines the category of 'too young to be held responsible'?

In similar vein, let's consider the ongoing debate around assisted suicide. There are two categories lurking within that phrase: 'assisted' and 'suicide'. But what, precisely, constitutes the former? If you were to beg me to help you end your life and I were to inject you with a drug that instantaneously kills you, then yes, that would definitely count as assistance. But what if I were simply to purchase your ticket to Geneva? Or give you a lift to the airport? Can that be defined as assistance? If not, can it nevertheless be construed as the thin end of an ever-expanding wedge? Where would Eubulides' elusive heap of sand start forming in this instance?

The exercise isn't a game. It's a serious cognitive workout, the object of which is twofold. The sharpening of one's logical reasoning skills. And the improvement of one's clarity of thought. Identify on which side of the fence your line of argument falls, and on which side of the fence opposing arguments fall, and you'll be in a better position to know what category to put them in.

More importantly perhaps, what label to put on the side.

Step 2. Once you've selected the appropriate categorical units in which to express your argument, make sure they adhere to the five components of the SPICE model.

Gavin Hewitt, the BBC's former Washington correspondent, was part of the media entourage that shadowed Donald Trump during his 2016 election campaign. As the months wore on, he got to know the man a little. On one occasion, Gavin told me, at some point or other during the campaign, Trump tweeted a picture of himself sitting aboard his private jet with a family-sized bucket of KFC into which he tucked with a silver knife and fork. The press, a good portion of it at least, tore him apart. The cutlery was one thing. But eww, fried chicken?

According to Gavin, however, the tweet was a stroke of genius. Trump, he pointed out, isn't the slightest bit interested in appealing to everyone. Never has been. He *wants* to create binary divisions. His audience is made up of white, working-class and non-college-educated folk who feel completely overlooked and blithely disenfranchised by the liberal, intellectual left. As long as there are enough of them, and they vote for him, it's game over.*

---

* At no time, perhaps, in Donald Trump's presidency has his divisiveness been in greater evidence than when his defence secretary, Mark Esper, implored

But *why* was that tweet so powerful? What was it in particular that made it into such a statement? The answer, it turns out, is that it packed a lot of SPICE.

Simplicity: What could be more graphic and 'in your face' than a bucket of KFC on a private jet?

Perceived Self-Interest: There's nothing better than fried chicken, right? Tuck in!

Incongruity: KFC . . . Silver service . . . The White House?

Confidence: Hey, why pretend? I may be loaded and running for office but I do like my chicken nuggets.

Empathy: I'm a regular guy who gets you and wants to represent you. Hillary's got a private chef.

---

governors to dramatically ramp up the use of the National Guard to 'dominate the battle space' against protestors in the highly sensitive aftermath of the death of George Floyd, a 59-year-old black man who lost his life while pinned to the ground by four white law enforcement officers in the Powderhorn neighbourhood of Minneapolis on 25 May 2020. Handcuffed and lying face down in the street in front of a gathering crowd of bystanders, Floyd repeatedly told one of the officers, 'I can't breathe', pleading with him to remove his knee from his neck. The officer, Derek Chauvin, failed to comply, prompting worldwide condemnation and mass rallies in a number of major cities – London, New York, Athens and Sydney among them – against institutional racism. Chauvin was subsequently charged with second-degree murder while the other officers present during the arrest were commensurately charged with aiding and abetting. A week later, with government officials forcibly clearing demonstrators from his path, Trump – who had previously upbraided governors for being too passive in the face of the ensuing civil unrest, urging them to 'dominate or you'll look like a bunch of jerks' – walked through a park in Washington, DC to stand outside a boarded-up church holding a Bible aloft in his right hand. With an election looming later in the year, the president's dictatorial pledges to 'dominate the streets' and his implacable self-portrayal as the 'president of law and order' appealed to many of his supporters, as did his use of religious imagery. 'Every believer I talked to certainly appreciates what the president did and the message he was sending,' said Robert Jeffress, the pastor of First Baptist Dallas and Trump stalwart. 'I think it will be one of those historic moments in his presidency, especially when set against the backdrop of nights of violence throughout our country.' Millions across the world saw fit to disagree, however: the family and friends of George Floyd first and foremost among them.

But there's more. Trump's tweet also aligns with our three binary super-categories to turbocharge his message. *Fight versus Flight*: Here's who I am. If you don't like it, suck it up! *Us versus Them*: Are you gonna eat fried chicken or . . . *tofu*? *Right versus Wrong*: Don't put on an act. Be true to who you are. The fried-chicken-eaters will inherit the earth.

To hit those persuasion hot buttons you first need to SPICE up your message.

Step 3. Use metaphors to ensure that your ideas travel around the brains of those you are trying to influence using the path of least resistance.

It can feel clunky and artificial to think in terms of images, analogies and allegories at first. But if you practise exporting your thoughts into a figurative, metaphorical format it won't be long before you start to reap the rewards. Not only will it start to feel more natural, you'll also become a better persuader. Language, as we saw in Chapters 7 and 8, can induce powerful psychoactive effects – and when cut with simile and metaphor such effects are particularly potent.

In my capacity as a performance consultant in elite-level sport I work with some of the UK's top track and field endurance athletes. Psychology is important in athletics, as it is in any sport, because when you're as honed to physical perfection as a potential Olympian, it's what goes on upstairs that invariably tips the balance. Ninety-five per cent of it is psychological, I tell them. The rest is in your head!

For months I'd been trying to figure out a way of impressing upon some of my protégés the need to distribute their effort evenly throughout a race; how not to go out too quickly and blow up early, but not to start off too slowly and leave it all on the track. Nothing seemed to be working. I just wasn't getting through. Then one day I hit upon the analogy of money.

'Think of energy as cash in your pocket,' I explained. 'When you go out for an evening you don't want to crash and burn before you've had a good time, do you? How are you going to feel when you have to go home skint while your mates are still out on the town enjoying themselves? Spend it wisely, pace yourself and you'll still be going strong at the end of the night.'

Mercifully, this did the job. So well, in fact, that it's now become part of the banter even when I'm not around. Why? Well, while these guys aren't exactly rolling in it, some of them do enjoy the benefits of corporate sponsorship and lottery funding and, for the first time in their careers, have some cash to splash on the occasional little jolly. They could identify with the narrative. It meant something to them. The metaphor made it real.

Another thing you find with elite-level sportspeople is that they'd rather go to hell and back than admit to their coaches they're injured. No one thinks this is a good idea. Not even the athletes themselves. But still they persist with this self-defeating strategy because when the plane takes off – for the World Championships, for the Olympics, for the European Championships – everyone wants to be on it. The irony, of course, is that the more you keep quiet and say nothing, the more likely it is you'll be grounded: that you'll be waving a farewell handkerchief at the airport and then blubbing into it all the way to rehab. Training through an injury inevitably makes it worse and nearly always ends in tears. How do you tackle *that*?

Precisely because this is one of those situations where there is no universal antidote it's also one where the clever use of metaphor really proves its worth. One of the athletes I work with is a Formula 1 fan. So what better analogy could there possibly be than that of coming in for a pit stop?

'Imagine you're a driver and you know you need to pull your car into the pit,' I told him, 'but you keep putting it off because you don't want to lose valuable seconds. What's going to happen?

Eventually the car will blow up completely and you'll be out of the race for good. You won't even reach the finish line, let alone be swigging Champagne on the podium.

'On the other hand,' I continued, 'if you get your car checked out, as all drivers have to do at some point, then you might lose a bit of time while you're being seen to but at least you're still on the circuit and in the mix.'

Almost overnight, the athlete's mindset changed dramatically. A chequered history turned into a chequered flag.

You can practise Steps 2 and 3 by asking friends to toss some arguments at you while you select the appropriate categories in which to frame them and the suitable kind of language in which to express them. But whatever cunning metaphor you eventually decide to go with, it can't just be something you've plucked out of the air at random. First and foremost, it needs to be relevant to whoever it is that you're talking to. The athletes loved the money and car-racing analogies because both, in turn, spoke to their interests and outlooks. They identified with them. Think of it in terms of nailing the right frequency – the appropriate psychological bandwidth – on which to transmit your message.

Several years ago a group of researchers in the US demonstrated this in a study. They selected a bunch of messages containing sports metaphors (such as, 'If college students want to *play ball* with the best, they shouldn't miss out on this opportunity') and compared them with a bunch of neutral messages ('If college students want to *work* with the best, they shouldn't miss out on this opportunity'). Which of these two message types, they wanted to know, would create the greater interest? In addition, which would the students consider the more persuasive?

The answers were as they suspected. Analysis revealed that the messages incorporating the sports metaphors were not only given more thought, they also had greater impact. But only for the students who were sports fans. For those who weren't interested in sport, the imagery proved counterproductive. It resulted in less persuasion and interest.

This is a truth that the New Zealand Prime Minister Jacinda Ardern grasped brilliantly during the early days of the COVID-19 crisis. Following a special Cabinet meeting convened in Auckland on 14 March she announced at a news conference that New Zealand needed to 'go hard and go early' in order to flatten the curve of new cases of the virus and avoid the healthcare system being overwhelmed. Though Ardern didn't explicitly state as much at the time it is highly unlikely that her words weren't carefully chosen to resonate with the sporting sensibilities of the rugby-obsessed nation. As Alastair Campbell later pointed out in an article in the *Independent*, those five key words that comprised her central message 'sounded like something from an All Blacks team talk'.

The figures speak for themselves. At the time of writing, New Zealand has recorded only twelve deaths from Coronavirus. The UK, 16,509.

Metaphors, it would seem, are like dentures. They're only any good if they fit.

Step 4. Next time you read a news report or see one on TV, pay particular attention to the kind of picture the influencer is trying to paint in order to get you to see things their way. Or, more specifically, identify the category or categories they're using to frame their argument and then see if you can outsmart them by considering if there are other arguments and categories that might better fit the bill.

Sprinter Christian Coleman learned this lesson the hard way. When he bagged the 100-metre gold in 9.76 seconds at the 2019 World Championships in Doha, making him the sixth-fastest man in history, he'd barely made the start line. He'd previously been charged with a whereabouts violation by the US Anti-Doping Association that, had it not been thrown out on a technicality, would have resulted in a two-year ban. His offence? Failing to log his physical location on a given day and at a given time on three separate occasions during the preceding year.

The testing regime works like this. Every day, an athlete is required to update an app on their phone to say where they will be for a specific hour of that day, so if the testers decide to randomly check their urine they know where to come and find them. It's a pain, for sure. But given the low level of public trust in world athletics at the time of writing, a necessary one. What interests us here, of course, is not what Coleman did but what he *said*; how he chose to either exonerate himself if he thought he was innocent or put his hands up if he thought he'd been negligent.

He elected to do the former.

'I didn't do anything wrong,' he observed. 'I haven't been careless. Everyone in this room has not been perfect. I am just a young black man living my dream, but people are trying to smear my reputation.'

Notice the use of the three persuasion super-frames here.

Fight versus Flight: The athlete chooses to defend his actions (or lack of them).

Us versus Them: To him, this is about young black men against the racist media.

Right versus Wrong: You're clearly out of order if you're racially prejudiced.

But Coleman had a problem. He'd hit all three buttons on the supersuasion mixing desk all right. Trouble was, he was mixing the wrong track. He'd plumped for the wrong persuasion category. To most people, it transpired, including other black athletes, the issue was nothing about race. Nothing about colour. Instead, it was about missed drug checks that were vital to the integrity of his sport. And about him being *world champion*. Whether he was a young or old, black or white world champion didn't matter. What *did* matter was that he was now a high-profile role model for his sport and should be seen to act as one. He was widely slammed in the media.

Coleman's case is among a number of high-profile 'category wars' to have featured in the news of late. Billionaire entrepreneur Elon Musk's allusion to Vernon Unsworth, a British diver who was instrumental in coordinating the Tham Luang cave rescue of twelve boys and their soccer coach in 2018, as 'pedo guy' on Twitter was framed by *his* lawyers as an insult and by Unsworth's as an accusation.

Insult won the day.

In a recent address to the Anti-Defamation League, the actor and comedian Sacha Baron Cohen pointedly distinguished between 'freedom of speech and freedom of reach' when referring to Facebook's enduring resistance to removing political ads and other social media posts containing hate speech, intentional lies and other forms of gratuitous misinformation from its platform.

The jury's still out on that one.

So, how might Coleman have run the race differently? What other category or categories might have better suited his argument? Well, how about that of 'immature young man makes a mistake and vows to do better in future' for a start? In other words, the category of inexperience. That might have played out like this.

Fight versus Flight: OK, I'll stand up to be counted and face the music.

Us versus Them: OK, I was foolish and immature. But hey, who hasn't been?

Right versus Wrong: OK, You're right to pull me up on this. But you'd also be right to give me another chance.

As it turned out, all charges against Coleman were eventually dropped and the athlete was exonerated from any wrongdoing. It doesn't change a thing.

Choose your words wisely.

And your categories even more so.

Step 5. When persuading others, or when others are persuading you, always come back to the three binary categories of supersuasion: Fight versus Flight; Us versus Them; Right versus Wrong.

Imagine yourself alone in a park in Belfast, circa 1985.* It's Halloween, two in the morning, and you can see your breath in the icy, sodium-lit air. Great. If you can see it, then others might see it too, and when you're staking out the house of a leading Republican dissident you can do without unwanted social calls.

A rustle pierces the silence. In bushes a few metres away to your left, something stirs. Calmly, casually – it's all about blending in – you give it the once-over. Nothing. Probably a fox, you think. But there's no mistaking the voice. Then menacingly, out of the darkness, there they are. Just ten or so feet in front of you on the grass. Parkas, tracksuits, bomber jackets . . . a gang of youths out for an evening constitutional. A gentle spot of late-night trick or treating.

You know what they're thinking. They're thinking you're one of 'them'. The filth. The fuzz. The enemy. Although you've been trained to within an inch of your life to deal with pretty much anything, this is when that inch comes in handy. There are more of these lads than you. A lot more. What's worse is they're out of booze. Not good. Empty bottles of Jameson have a nasty habit of getting shoved in people's faces in these parts.

This could get a little bit interesting.

'Fair play to you, fellas,' you say, pulling out a pack of Bensons

---

* The Troubles, also known as the Northern Ireland conflict, was a violent sectarian struggle in Northern Ireland marked by various paramilitary activities that began in the late 1960s and is generally considered to have been brought to a close in 1998 by the Good Friday Agreement. Political and nationalistic in nature, the key issue behind the conflict lay between the Protestant unionists – or loyalists – who wanted the province to remain within the United Kingdom, and the Catholic nationalists – or republicans – whose agenda demanded that Northern Ireland leave the United Kingdom and become part of a new, united, Republic of Ireland.

from an inside pocket and offering it round. 'You got me, I *am* a cop. But look, it's not you I'm interested in. Truth is, I'm on an undercover job trying to track down a paedophile. He's preying on kids around here.'

No one moves a muscle. You could hear a pin drop. Then suddenly, imperceptibly, expressions change. Body language slowly thaws. And the standoff quickly resolves. The lads turn to each other and nod. And silently they melt away.

It's the kind of thing that only happens in the movies, right? Wrong. This is a true story told to me by a real-life member of the British security forces I interviewed a number of years ago. He had one chance, one go, one shot at getting out of that situation. Remarkably, it hit the target.

How?

Supersuasion!

Fight versus Flight: Is that what you want . . . paedophiles around here preying on kids?

Us versus Them: Either you help me or you help the paedo. Which is it?

Right versus Wrong: Does screwing up an operation against a child rapist make you feel good?

It's simple. The greater your awareness of supersuasive principles – how what you say and how what others say to you matches up with these three influence super-categories – the better a persuader and the more persuasion-resistant you'll become.

Let's finish off by putting in another call for some Kentucky Fried Chicken. Only this time without the private jet and silver service. Several years ago in the UK KFC famously switched its delivery contract to a new provider and ended up succeeding, in swift and spectacular fashion, in exhausting its entire supply of chicken in over 250 outlets across the country. For a company that pretty much just sells chicken that's one hell of a key ingredient to be running out of. What the cluck? More to the point, what the cluck could they do about it?

The answer, it transpired, was nothing short of a PR miracle. Over the chickenless course of the ensuing few weeks members of the public were, in varying degrees, both amused and surprised to notice a series of prominent ads appear in the *Sun* and *Metro* newspapers featuring Colonel Sanders' iconic little red and white bucket.

Amused? Surprised? What was going on? Well there was, it turned out, something just that little bit different about this particular bucket. You see, the receptacle in question was emblazoned not with the initials KFC.

But FCK.

And underneath were the words: 'We're sorry.'

Now there would have to be something seriously wrong with you to not see the funny side of that. KFC knew full well that its product was somewhat dressed down (even if it *was* endorsed by certain individuals aspiring to high office) and that its customer base would be particularly unenamoured with corporate-sounding, mealy-mouthed apologies. 'Mistakes were made' or 'We're sorry for any inconvenience caused' just weren't going to cut it. Instead, their approach appealed to our three persuasion super-categories while adding a pinch of SPICE.

Fight versus Flight: OK, busted! But we're not going down without a gag.

Us versus Them: We trust you, our customers, to get it. Everyone else can cluck off.

Right versus Wrong: We've taken a massive risk here reaching out to you like this. So how about giving us a second chance and coming back in to see us . . . ?

Simplicity: KFC to FCK? Not the most onerous design shift ever attempted.

Perceived Self-Interest: Have a joke on us.

Incongruity: Expletives in a nationwide ad campaign? You don't see that too often.

Confidence: People *are* going to talk about this.

Empathy: We know this is a bit of a gamble. But we reckon it'll raise more laughs than eyebrows. Haven't we all messed up?

Searching for nuggets in all of this (and who wasn't a couple of years ago)? Here's a big one. If you want to be a great persuader it never pays to be chicken.

CHAPTER 12

Redrawing the Lines

Did you know that the human voice is the only pure instrument? That it has notes no other instrument has? It's like being between the keys of a piano. The notes are there, you can sing them, but they can't be found on any instrument. That's like me. I live in between this. I live in both worlds, the black and white world.

NINA SIMONE

IN JULY 2019, the British broadcaster, writer and former newspaper editor Piers Morgan fronted a TV show called *Psychopath* in which he interviewed a convicted killer behind bars at a supermax prison in America. At around the same time the show was broadcast in the UK, to complement it, I published a piece in the *Radio Times* TV listings magazine in which I assessed where Piers himself sat on the 'psychopathic spectrum'. To do this, I presented him with a dozen or so psychometrically validated statements to which he responded via a four-point scale according to how accurate, or inaccurate, a description of himself he considered each one to be.

The results were as I'd suspected. No Hannibal Lecter. But no

Dan Walker either.* We settled on the diagnosis of 'good psychopath', the term I use to describe individuals who are able to dial up their inner ruthlessness when the situation requires it. And similarly, when needed, dial *down* their conscience and empathy. It's the title of a book I wrote a few years ago with the former special forces soldier Andy McNab on how psychopathic characteristics can be beneficial if used in the right context, with the right intentions, and deployed in the right combinations and at the right levels. That book caused a lot of trouble.

Piers likes a bit of trouble so I tell him about it. It began before we'd even put pen to paper. Companies House banned us from trading under the name 'Good Psychopath Ltd' because they deemed it in poor taste. We appealed the decision. And won. Then there was the fan mail. Some of it good. Some, well, not so good. A couple of years after the book came out, for instance, I wrote a cover feature for *Scientific American Mind* in which I evaluated the four US electoral candidates, Donald Trump, Hillary Clinton, Bernie Sanders and Ted Cruz, determining where each of them sat on the psychopathic spectrum, and, as a postscript, where each of them figured in a psychopath 'league table' of famous historical personages.

The Oxford University switchboard jammed, the media went nuts, and I received a slew of entertaining emails from mildly disaffected Trump supporters, one of which was accompanied by a picture of me in a gas chamber with Donald flicking the switch.†

---

* 'Saint' Daniel Walker, as Piers Morgan sometimes calls him on the ITV breakfast show *Good Morning Britain* that he co-hosts with fellow journalist Susannah Reid, is Piers' super-chipper opposite number on the competing BBC show *Breakfast*. The two share a (largely!) good-natured rivalry.
† The irony is that, far from being derogatory about Trump, my article put forward the scientific argument that certain psychopathic personality characteristics comprise the necessary criteria for success in certain high-stakes, high-pressure professions, politics being one of them.

It concluded with the following words:

It will be my pleasure to ensure that you are banned from entry to the USA. I am sure a further investigation will find criminal conduct on your part. So the next time you are sunning yourself on a beach in Portugal don't be surprised if US government agents drag you by your balls (if you have any) to a special flight.

I understand the state of Louisiana maintains some wonderful facilities for treating mental disorders such as yours. Five or six years learning the finer points of picking cotton should help you immensely! A steady diet of cornmeal mush and dirty water should make you fit as a fiddle.

Waiting to greet you upon your arrival in America,
Wayne

The picture is up on my door.

Donald, needless to say, was extremely high – on both spectrum and table – prompting a slew of now infamous headlines, best summed up by this black-and-white hall-of-famer in the UK's *Daily Mirror*: 'Psycho Don Trumps Hitler'.

Piers chortles. He used to be their editor.

'I've thought about this,' he says. 'And it seems to me that we've gone full circle. Millions of years ago when our cave-dwelling ancestors lived in small, close-knit tribes they used to run around with lumps of rock attacking and killing each other if those tribes got into a spat. Then presumably, at some point, we developed the higher-order faculties of language and reason to help us sort out our disagreements in a more constructive, less hostile fashion.

'But today, on social media, it seems we've gone back to the bad old days, to the prehistoric way of doing things. We're back to instinctive, kneejerk, all-out attack mode. You've just got to check out my Twitter feed. Yours, too, I bet. It's ridiculous. And, to be honest, it's all just a little bit depressing.

'If someone from a tribe over here says something that someone from a tribe over there doesn't like, what do they do? They click-and-run. They come piling out of their online echo-chamber caves with their chat-room coshes and their keyword cudgels and steam into them.

'Has natural selection been wasting its time over the last few million years? Has it been barking up the wrong tree . . . or *down* it, should I say? In this current, right-on era of virtue signalling, vitriol and victimhood, has evolution sold us a dummy with our big, high-definition, widescreen brains? Should we not, in hindsight, have gone for a cheaper model?'

Piers is spot on. Right now, 'us and them' – 'clique-baiting', you might call it – has never been bigger business. Social media may have its problems, but if there's one thing it *is* good at it's bringing like-minded kinsfolk together. So much so that the sprawling electronic savannah has become politically and ideologically overrun. Not, as Piers points out, with friendly, peaceful and cooperative tribes. But with hostile, hawkish and warlike ones. It has become dangerously overpopulated with all sorts of marauding online clans – forums, platforms, communities, hashtags – that troll first and then ask questions later. Anonymity fuels aggression; and constraints on time and maximum word and character limits make nuance a luxury that is now no longer affordable. They heighten, as we learned from Arie Kruglanski, our need for cognitive closure which brings down the curtain on impartial, informed analysis halfway through the show. Degrees of disagreement have ceased to be legal tender. The currency is truth. Absolute, incontestable truth. And there's just one denomination. Mine.

The price on dissent is considerable. Mess with people's categories, devalue this currency of militant, subjective absolutism, as I discovered to my cost, and you're going to get into hot water. You're going to be painting a target on your back. Until my book on psychopaths came along, everyone was perfectly happy with the concept. There were millions of other books out there on gang-sters, rapists and serial killers telling us how evil they were. And no one batted an eye. They still don't. Our brains construct a morbid mental menagerie in which the monsters and bogeymen of our deepest, darkest nightmares are all held securely under taxonomic lock and key, responsibly banged up in their escape-proof categor-ical cages, allowing the rest of us to circulate and observe them at our leisure, safe in the knowledge that interposed between us and them stand six-inch bars of reinforced cognitive steel.

But *The Wisdom of Psychopaths* changed all that. It opened the doors of those carefully constructed cages and let the bogeymen and monsters roam free so that all of a sudden, quietly and casually, they were 'out' and walking among us.

At the flick of a categorical switch, that buffer of 'us and them' previously installed by those rickety criminological enclosures of clinical diagnosis and forensic classification smartly shot up, like the shutter on a ticket-office window, and the race was on to manhandle it back into place. Like many revolutionary ideas, the assertion that not all psychopaths are bad and that some of them are good – that, just as the rest of us are not always 'right', psychopaths, correspond-ingly, are not always 'wrong' – challenged our perspective on who we really are. It took that neat, orderly row of psychological crayons that natural selection has painstakingly arranged for us on our brain's pattern-recognition desktops, and which we use with unstinting disregard for subtlety and shade and nuance to colour in the outline of everyday life and experience, and ruffled them all up. It effaced the lines along which we think and live; those three, all-embracing, super-categorical lines that since the earliest stirrings of the dawn of prehistory have formed the key developmental axes of our tribal, tripartite brains.

Fight versus Flight: Do those who go where angels fear to tread have more in common with the demons that they rub shoulders with than we might otherwise have thought?

Us versus Them: Who among us can cast the first stone?

Right versus Wrong: Might virtues and vices be cut from the same cloth?

The lines redrawn, it would take some getting used to.

Facts versus truth

We've come a long way since we were kicking around on the Russia–Finland border a century or so ago at the beginning of Chapter 5. That line, you may recall, also needed redrawing, and no sooner were the theodolites put away than Russian and Finnish winters suddenly became very different from one another, to the mind, at least, of a wily Finnish farmer. Of course, a matter of a few metres left or right, east or west, was never going to make any real difference to the isobars. And chances are that the farmer was as mindful of that fact as were the Russian and Finnish officials who looked in on him that day. But there's also a chance that it wasn't *all* Machiavellian, self-interested sophistry on his part and that no sooner had our beleaguered frontiersman begun thinking in terms of a black-and-white border than a genuine misperception of conditions either side of it began to set in. Drawing lines doesn't just serve to divide. It also conspires to *exacerbate* division. Between people, concepts, creeds, philosophies, actions . . . between whatever it is that we perceive as sitting on 'this side of the fence' and on the other, across the hair-trigger semantic trip wires of salient categorical boundaries. Construct a cage for any animal or creature, or as we've traditionally done with psychopaths, place a putative 'beast' or 'monster' behind bars, and they instantly become more dangerous, if not in practice then certainly in the realms of our febrile imaginations. And there's some pretty good science to prove it.

Take colour. Studies show that when making similarity judgments between three different shades, observers appraise two shades from the same linguistic tribe (e.g. A and B in Figure 12.1

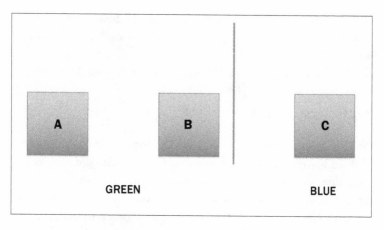

Figure 12.1: Colour Prejudice.

here) as being closer in hue than two shades from different linguistic tribes (e.g. B and C), even though the perceptual distances
between each pair (in terms of electromagnetic wavelengths) are
identical.

The simple fact that Hue C is adjudged to be a chromatic citizen of Colour Tribe BLUE is sufficient to render it – subjectively,
at least – as sharing less electromagnetic DNA with Hue B than
does Hue A, a fellow citizen of GREEN.

To grasp how the *principle of accentuation* (as it's known to experts
in the field of social cognition) might work in more everyday life
settings, let's switch our attention from the chromatic to the kinetic
and entertain a little thought experiment. Imagine that you have to
evaluate the top speeds of three cars: A, B and C. You drive the cars
around a racetrack, flooring the accelerator on each circuit, and then
come up with an estimate (the speedometer, needless to say, is dis-
connected). But there's a catch. Car A is a Ferrari. Car B is a Maserati.
And Car C is a Fiat. Plus another catch. I've rigged each vehicle
with a speed limiter. The maximum velocity of the Ferrari is 90mph,
the Maserati 80mph, and the Fiat 70mph.

How do you think you'd get on? There's no way of telling until you give it a shot, is there? But one thing you *can* be certain of is this: you'll judge the top speeds of Car A (Ferrari) and Car B (Maserati) to be more similar than the top speeds of Cars B (Maserati) and C (Fiat), even though, just as in the colour example, the difference in velocity between each pair is actually the same (10mph).

The reason for this is simple. Everything else being equal, you might begin the task with the reasonable assumption that cars that are more expensive are also faster. So, when asked to estimate the speed of a car (its 'focal' attribute: the feature you're being asked to evaluate), your assessment of that attribute is likely to be contaminated by what make it is and how far along the prestige spectrum (or 'peripheral' dimension) that make might happen to fall. Is it a sports car, like a Ferrari or a Maserati, or a family car like a Fiat? The further along, the faster, by implication, the car.

In other words, as soon as we introduce the notion of entities falling consistently into different, distinct *categories* – nationalities, colours, prestige brands, gender, psychopath, etc. – our appraisal of those entities and their various focal attributes will no longer be the result of pure, unfiltered perception, but rather will be influenced by the higher-order consideration of category *membership*: the cognitive 'periphera' of us, them, this, that, the other.

No sooner do we talk about categories than we lapse into black and white.

A great example of this recently played out on the political stage in the UK. In March 2020, in an attempt to slow the spread of the Coronavirus pandemic that at the time was gathering momentum, the former British Prime Minister Tony Blair suggested in an interview with Sky News that mass testing of a significant proportion of the UK population – 'virtually everybody' – was an essential step in the fightback against the disease. Reaction on Twitter was about as polarized as it could get. Some thought it an eminently sensible suggestion. Others, many of whom also took the opportunity to denounce Blair as a 'war criminal' for failing, as they saw it, to conduct due diligence before committing the lives of British service personnel to an 'unjust' and 'unlawful' campaign in Iraq in 2003, deemed it absurd, with one even vowing to set up a petition to forbid him from issuing any further statements on the subject.

With opinion so split down the middle, it's easy to see what happened. The 'peripheral' criteria of political allegiance and disaffection over Iraq instantly stole the limelight from the 'focal' issue of combating COVID-19. With attention firmly focused on Blair as opposed to the virus his comments were judged not on their content, or their wisdom, or their feasibility. But on their source.*

A week or so after the Sky News interview, on the day that Sir Keir Starmer became the new leader of the Labour Party, I put it to Tony that such has always been the case in politics. Rather than turn a tap on to fill a glass some people opt for a water cannon. His words, frank but philosophical, offer a thumbnail of the book as a whole.

'There has always been a high level of criticism in politics and often abuse of those with whom we disagree,' he told me. 'Politics is controversial. Government is unbelievably tough. When you decide you divide. Pleasing all of the people all of the time is neither possible nor desirable. So none of this is new. Just look at the abuse directed at Abraham Lincoln or Winston Churchill, at least in the 1920s and 30s.

'What *is* new is a culture in which the decline of deference and the rise of the media mean that scrutiny, fair and unfair, is much more savage and less understanding of the real challenges of political life; and above all, we now have the advent of social media.

---

* For a more detailed explanation of the possible factors contributing to some people's seemingly disproportionately negative reaction to Blair's remarks, see Appendix VII.

This literally transforms the context in which criticism – and attendant abuse – occurs. Everyone has an opinion. Everyone now has a platform to express it. It works through impact. The more extreme, the more black and white, the more impactful. Even if it's less rational.

'The one piece of good news is that the public, despite it all, does understand that this new world isn't a guide to the real world. The politician or indeed anyone in the public eye has to tolerate what really shouldn't be tolerated. But if you're able to do that, then there is still a market for reasonable debate and the exchange of views.

'Most people are polite about disagreement. They don't set out to polarize. Those who do will achieve the publicity they crave. But the silent majority of the reasonable are the best antidote. And you simply have to have the character and resilience to know that and overcome the battering by not dwelling on it!

'But that's easy to say and very hard to do!'

Of course, the strange magic of accentuation has not been lost on those who work in the advertising industry. Brand establishment, and subsequent brand management, is key to market success. A good brand isn't just a label. It's an experience. One which sometimes affects us without us even knowing. Back to colour, for instance, and several years ago a US market research organization experimented with different shades of background on cans of 7-Up. Some cans were more yellow. Others more green. It wasn't long before the phone lines at head office began to light up. Those who purchased the yellow cans reported an unfamiliar 'lemony' flavour to the contents while those who bought the green cans complained of too much lime in the mix.

But it's when we come on to people and the effects that categorization and accentuation exert on the relationships that we have with each other that the storm clouds begin to gather. Because irrespective of the nature of the *focal* dimension upon which we might formulate our judgments of others – a particular aptitude or ability or skillset, for instance – the *peripheral* criteria of category membership – socioeconomic status, religious inclination, political affiliation, sexual orientation, gender, race, you name it – become entangled around the roots of our socio-cognitive processes like psychological knotweed. We assume. Jump to conclusions. Fill in the gaps. We derive, exactly as evolution intended, maximum information from minimum cognitive effort.

In short, we begin to stereotype.

This is hardly surprising. Ever since big brains started taking over from big muscles, fast legs and sharp teeth, natural selection – as we saw in Chapter 10 – made the eternally binding decision to compel us to hang out in groups. And no sooner did this propensity to band together become superimposed upon our hardwired capacity to categorize, to divide the world up into bite-size chunks of optimally encodable and maximally informative data, than we began asking for serious trouble. *Us versus Them* became a proxy for *Good versus Bad*.

Not only did we start to stereotype, we also started to denigrate.

Just how fastidiously natural selection managed to install our tribal circuitry, and just how easily and effortlessly its skullduggerous switches are flipped, we saw when we watched the Dartmouth Indians and Princeton Tigers kick ten shades out of each other back in the fifties. In the aftermath we learned something else: how 'what *we* do is right and what *they* do is wrong' comprises such a powerful moral heuristic that it doesn't just influence what we think. But also what we see.

Evidence of this link between the collective and corrective is now well established, so much so that experts working in the field of belief formation – those who study the precise cognitive and psychological means by which we come by our attitudes and opinions – unanimously accept that we consistently perceive the in-group with whom we identify not only as being nicer, warmer and generally more agreeable than those we regard as different, but also sharper, shrewder and having a firmer grip on reality. We don't, in other words, just like them more. We trust them more too.

A number of years ago back in the early nineties, Dominic Abrams, professor of social psychology and director of the Centre for the Study of Group Processes at the University of Kent, performed a fiendishly elegant experiment which captured this phenomenon beautifully. Abrams began by presenting a classic optical illusion, the autokinetic effect, to a group of six participants in a blacked-out, pitch-dark room. The illusion isn't complicated but it's powerful. For a period of about fifteen seconds, a stationary point of light appears to oscillate randomly in front of you against a backdrop of total darkness. The participants' job was easy. Over a series of trials they had to estimate, out loud, the farthest distance they thought the spotlight had travelled relative to its original position.

But there was a catch.

Half of the group were accomplices. They'd been briefed by Abrams to extend the judgments of real participants by five centimetres. And that wasn't all. In addition, Abrams also subtly manipulated the social identities of some of these secret agents so that they bore either greater or lesser resemblance to those of the genuine participants.

The question he was asking was simple. Would the pull of group 'belongingness' cause participants to increase their estimates to complement those of the infiltrators that they believed were most like *them*?

The answer was equally simple.

Yes.

Results revealed that the greater the perceived disparity between the participants and the decoys – the more they appeared to belong to different social groups – the greater the discrepancy in estimates. In contrast, the greater the perceived similarity, the closer the alignment in judgments.

The take-home message was clear. Those who are with us are right. And those who are against us are wrong. Not only that, but those who are 'usser' are righter. And the 'themmer' you are, the more wide of the mark you're inevitably likely to fall.

Not the end of the world

'There are the facts and then there is the truth,' an old teacher of mine used to say.

I once asked him what the difference was.

'It's a fact that a tomato is a fruit,' he said. 'But the truth is you wouldn't put it in a fruit salad.'

If you're into inspirational quotes and online motivational mood boards you may well be under the impression that in the existential poker game of life a pair of truths will beat a flush of facts. But in light of Abrams' *son et lumière* in the darkroom, you may want to reconsider. The study creates an hallucinatory subjective reality chamber in which the figments of real and unreal become seamlessly enmeshed in a chimeric luminescent judgment-fest of overlapping and competing identities.

What I see . . . is who I am . . . is who we are.

The *fact* of the matter was that the spotlight didn't move. Any change in its position was merely a visual illusion. Motion perception must always occur relative to a fixed reference point and, as all such points were automatically occluded by darkness, so the three-dimensional locus of any given item at any particular time must necessarily have remained undefined.

But the *truths* were a different story. The spotlight meandered about. Sometimes – Truth 1 – quite a lot, for those participants who identified with the magician's apprentices. Sometimes – Truth 2 – not so much, for those who had less in common.

There can only be one set of facts. But there can be any number of truths. Would *you* put tomatoes in a fruit salad? Me neither. But I bet we all know someone who would.

Now, for those of you who are one step ahead of me there exists a major, obvious and rather important difference between this tightly controlled, carefully orchestrated study and real life. And this is where things start to get interesting. In real life we're exposed to the facts. Having made our decisions and committed ourselves to the outcomes we have the opportunity to rethink our positions, to re-evaluate and change our minds.

But in the study this wasn't the case. Abrams' primary focus was on subjective variations in *truth* – in how *far* the spotlight had travelled from its initial point of origin – not on subjective variations in *fact*: whether it had moved at all. Prior to voicing their estimates, none of the participants had the faintest idea that nothing had actually happened; that the light had in *fact* remained stationary. Nor, once the experiment was over, were they afforded any opportunity to reconsider their responses in view of this simple detail.

The study was all about truth. Not about fact and how fact might impact on truth. But on truth in the absence of fact. Real life is all about both.

Of course, changing one's position in the face of new information might seem like the smart way to go. But often we're just not that smart. Often we go a different, slower, more congested and circuitous route. We double down; dig in our heels; clutch our convictions tightly to our chests like kids in a toyshop refusing to let go of some sudden, must-have plaything that Mum and Dad have told us to put back on the shelf. When sentiments pass their sell-by date the best thing to do is to scrape them into the bin. Replenish our belief racks with fresh, healthy, cognitively nutritious produce. But we don't. Not as regularly as we should do, anyway. Instead we store them in jars and keep them at the back of our thinking cupboards where they get in the way and take up valuable reasoning space.

We pickle and preserve them in truth.

On the subject of sell-by dates, here's how, and why, it works. Back in the 1950s the American social psychologist Leon Festinger became intrigued by an item that he'd spotted in his local newspaper. A Chicago-based doomsday cult calling themselves the Seekers had assembled a prophecy of highly inadvisable precision. As with all such exhibits of merciful, tender-hearted apocalyptica, the prediction was intriguing if a tad self-righteous and smug. At the stroke of midnight on 21 December 1954, the chosen few had illuminated, the world would be destroyed by a divine cataclysmic deluge of which *they* would be the only survivors, spirited away to a secret cosmological safe house in a custom-built flying saucer.

Fair enough, on the one hand, Festinger and his colleagues surmised. It was perfectly possible that their own invitation had somehow got lost in the post. But what, on the other hand, if the utterly unthinkable happened.

Nothing.

If, ridiculous though it seemed, the prophecy failed to materialize?

To raise the stakes and make things a little more interesting – it was, after all, only the end of the world – Festinger, an expert on group dynamics, elected to proffer a startling divination of his own. If, at the appointed hour on 21 December the intergalactic cavalry didn't show, then allegiance to the cult, and support for the cult's leader – a Chicago housewife by the name of Dorothy Martin – would, rather than disappear down the plughole, leaving the rest of the cosmos safe and sound behind it, actually grow even stronger.

The reason for this, Festinger proposed, had to do with our need for authenticity, with our desire to be 'true to ourselves'. Or, more specifically, with our brain's evolved propensity to ensure that we act in a reasonable and rational manner commensurate with our beliefs: if we didn't, he pointed out, then we would fail to maintain a consistent sense of self and have little recourse to predicting the behaviour of others.

To this aim, he continued, whenever we are caught, for whatever reason, in a tricky situation in which our actions are at odds with our beliefs, the air in our brain becomes thick with psychological tension – an aversive computational weather front that Festinger termed *cognitive dissonance* – which we feel compelled to disperse through mental realignment: by redressing the balance and attempting to restore equilibrium between our attitudes and beliefs on the one hand and our divergent decisions and self-contradictory exertions on the other.

Sometimes, of course, we do this very simply through the act of changing our minds. Sometimes, if it's no big deal, we're perfectly happy to revisit what we've done, to revise our behaviour in the light of additional evidence. To illustrate, imagine that you buy a coat from a store, get it home and decide that on second thoughts you don't like it. The store in question operates a straightforward, no-nonsense refund policy. What do you do? Simple. You return the garment to the place of purchase and either choose to buy a different one or ask for your money back. Tension between fact/ behaviour (buying coat) and belief/attitude (don't like it) resolved.

Cognitive dissonance dispelled.

In a similar fashion, with an issue like Brexit, if it turned out that you voted Leave but that you weren't as resolutely committed to your views on, say, immigration as first it might have appeared and that you could, in fact, 'take it or leave it', then you may well have been persuaded by a version of the so-called 'control versus cut' argument* and, all else being equal, have decided in a second referendum, to vote Remain.

---

* I could, for instance, have pointed out (as the then Prime Minister David Cameron failed to do at the time) that both Boris Johnson and Michael Gove, the Vote Leave troop's two main protagonists, had taken great care throughout

But the problem, as Festinger deduced, is that it isn't always easy to remodel one's behaviour, either for reasons pertaining to simple practicality or for those relating to the often deep-rooted and frequently intractable motivations that might precede and underlie our actions. What happens then is interesting. With the opportunity to 'turn back the clock' or to 'wipe the slate clean' no longer an option, rather than remodelling our *behaviour*, we redouble our *beliefs*. And instead of charting a different course, we justify the one we're on.

This is precisely the fate that befell the eponymous fox in Aesop's exquisite fable of *The Fox and the Grapes*, written some two thousand years before Festinger ran into the Seekers. It's a tale that could grace the pages of any psychology textbook. Plagued by hunger, a fox devotes considerable time and energy to snaffling a bunch of grapes hanging just out of reach on a vine above his head, yet try as he might, his efforts prove in vain. Skulking off in defeat, he convinces himself in hindsight that he doesn't really want them after all. 'Oh, you aren't even ripe yet!' he proclaims. 'I don't need any sour grapes.'

It's a familiar feeling. Imagine, for instance, that in the coat scenario it turns out that the store *doesn't* operate a simple and straightforward refund policy and you end up stuck with it. The only possibility available to you then to reconcile the fact that you

---

the campaign not to *attack* immigration or immigrants per se, but, rather, had maintained a singular focus on the ostensibly related yet, in truth, manifestly different issue of *controlling* immigration. Though it had gone largely unnoticed during the campaign itself I could've mentioned the fact – while at the same time making the case for the positive benefits of mass immigration for the UK – that neither Gove nor Johnson had at any point promised to *cut* immigration, but, instead, had talked only about an Australian-style, points-based immigration system and made a commitment to EU citizens living in the UK that their rights would be protected. Which meant, by implication, that those lawfully resident in the UK would automatically be granted indefinite leave to remain in the country with no less favourable treatment than before.

bought it with the belief/attitude that you no longer like it and eradicate the attendant dissonance is to somehow change how you feel about the garment. To 'come round' to liking it. To come to the conclusion that it isn't so bad after all. That it is, in fact, just the coat you've been looking for!

Similarly, in the Brexit example, if your views concerning the negative impact of immigration and the pernicious effects of EU membership on the UK generally were deeply entrenched, and if you were, let's say, well known among your peers for holding those views – indeed, you may have even canvassed on the issue in your local neighbourhood – then, once again, when confronted with an assembly of incontrovertible facts to the contrary, the easiest option available to you to preserve the alignment between your behaviour (supporting the Leave campaign) on the one hand and your beliefs (leaving the EU will enable the UK to cut immigration and 'take back control') on the other would have been to discard and discredit the newly acquired information while simultaneously reinforcing the position you currently held.

You might, for example, have pointed out that although those spearheading the Leave campaign had never explicitly pledged to cut the number of overseas nationals entering the UK, controlling the influx of migrants clearly amounted to the same thing, and that, by implication, the consensus was that 'mass immigration' posed a pernicious threat to the fabric of British society.

At the same time, to shore up your commitment to leave, you might also have considered flagging up a number of other perceived 'benefits' of Britain extricating itself from the EU: immunity, for instance, from some of the over-prescriptive legislature that, with unstinting pomposity and unfailing regularity, wafted across the Channel from Brussels ('Bananas should not be too bendy'; 'Prunes are not laxatives'; 'Eggs cannot be sold by the dozen') as well as the provision of greater economic freedom to make stronger and swifter trade deals with other nations.

The conundrum is as fiendish as it is absurd. The more committed one is to one's argument, the greater the psychological investment that one has in one's position, then the greater the probability that the provision of conflicting evidence will achieve the opposite effect to that which was originally intended: the higher the chance that far from instigating a hoped-for change of heart it will, in fact, harden that heart and data-slap the recipient into a posture of heightened intransigence.*

This, it turned out, just as Leon Festinger predicted, was precisely what happened with the Seekers who, under the savvy spiritual stewardship of Dorothy Martin, advanced the prognostication that the world would end through the divine upending of a giant cosmological ice bucket.

It didn't.

But did members of the cult retreat into the shadows in rancorous and repentant ignominy? No, they most certainly did not. And with good reason. The shadows that awaited them were very shadowy indeed. Many of them had sunk vast sums of money into the enterprise. Some had sold houses. Others handed in their notices in respectable, well-paid jobs. Then there were the broken relationships.

Dorothy Martin's acolytes simply had too much to lose by succumbing to the conclusion that they were, in fact, just flat-out wrong. So they didn't. Instead, the 'facts' were assimilated into an organic,

---

* I remember the actor and raconteur Peter Ustinov telling a story about a long-distance motor race – possibly the Paris to Dakar rally. Passing through one of the villages on the course one day a driver happened to run into a spectator, knocking him to the ground. Enraged, the villagers turned on the driver and killed him. Subsequently, when the injured spectator staggered groggily to his feet, they killed him, too, to justify their murder of the driver. While I am unable to swear as to the authenticity of the account, I include it, apocryphal or not, as an unparalleled example of the brain's insatiable need for cognitive consistency.

ongoing narrative that extemporized the fellowship's delusional beliefs and actions into a resplendently Machiavellian tapestry of divine intent. When the intergalactic Uber failed to pull up at the appointed eleventh hour to spirit them away from impending obliteration, Martin, in an uncannily timed development, received a message through automatic writing. Such was the light, keeping vigil through the night, that her little band had spread throughout the world, that God, in his mercy, had called the whole thing off. He had granted the earth a stay of execution.

Not only that, but in stark contrast to the strict media curfew that the group had imposed upon themselves in the immediate run-up to the anticipated cosmological catastrophe, once disaster had been averted and the Almighty had been appeased the order began proselytizing as if their very lives depended on it. Which, in many ways, they did.

Such was the dissonance between raw, objective, unrefined reality, and raw, subjective, misguided personal agency – and such was the magnitude of the financial, emotional and psychological commitment that the soothsayers had invested in their reveries – that the only way to get themselves back on track was to stay on the one they were on.

Identity squeeze

Ever since Leon Festinger first unveiled his theory of cognitive dissonance over half a century ago, sightings of it have been common. Hardened smokers dismiss nailed-down evidence that their habit comprises a leading cause of cancer as either 'ambiguous' or 'unproven', while dieters the world over tend to be renowned as much for the ingenuity of their get-out clauses ('It's good to cheat every now and then'; 'I'll do another thirty minutes in the gym'; 'The sugar content isn't *that* high . . .') as for their iron-willed resolve to eradicate the muffin tops and love handles.

Even Brexit gets in on the action. Six months after the UK elected to leave the European Union, CNN commissioned a poll of British voters to find out what would happen if the tape was run again and the EU referendum was reprised. The results were instructive to say the least, and bang in line with what Festinger might have predicted. Despite a slide in the value of the pound; despite accusations that the British government had failed to delineate a clear withdrawal plan; despite disparaging remarks from the so-called Remoaners that the UK had been swept out of Europe on a nationalist tide of paranoia and prattling, populist poppycock; and despite the fact that nearly half of the respondents were fully aware that the decision would hurt them financially – or rather, as Festinger would assert, precisely *because* of such contingencies – the Leave camp triumphed again. Of those questioned, 47 per cent said that they would vote to come out of the EU, 45 per cent said they would choose to stay in, while the remaining 8 per cent said they were undecided.*

If you can't turn back the clock,† then, as Dorothy Martin and her kooky cabal of divinatory desperados demonstrated quite admirably, sometimes you can't turn back.

Of late, however, things have started to change. Recently, we've begun witnessing a significant and conspicuous shift in the manner and means by which we deal with cognitive dissonance. Traditionally, as we've seen, when we're confronted by evidence contradictory to our beliefs and harbour a significant interest in maintaining those beliefs, sifting nuggets of truth from abundant seams of fact – interpreting the facts so that they align with what

---

* The result of the actual referendum in June 2016 was 52 per cent (Leave) and 48 per cent (Remain).
† A limitation which posed a significant confound to the CNN survey. The pollsters presented a hypothetical scenario, the responses to which would unquestionably have been influenced by the political status quo at the time; the situation, in other words, as it then stood.

we think – has been our go-to method for preserving a sense of identity, a strong and stable impression of who we 'really are'.

But over the past few years an alternative solution has steadily been gaining traction. Yes, as bulwarks of belief and guardians of value and principle, rebuttal, dissent and negation have always served us well. But shaking our heads and disaffirming the facts altogether marks a radical new departure in our pursuit of cognitive consistency, in our desire to preserve a clear, coherent and worthwhile sense of self. It heralds the dawn of a grave new world of partisan, 'post-truth' denialism. And with it, of course, the rise of 'fake news'. Reality as a public utility was, accordingly, once publicly owned. Facts were the same for all of us. But soundbite-by-soundbite, click-by-click, facts are being privatized. What's mine is yours and what's yours is mine . . . just so long as we're on the same side. Just so long as we're part of the same ideological tribe.

Perhaps unsurprisingly, given their occasional cameo appearances within the shadowy, labyrinthine plotlines of global governance and politics, there's been a fair amount written in recent times on the emergence of post-truth values.* High on the agenda of cautionary, blog-whistle commentaries we hear about the online fragmentation of traditionally more centralized, and ostensibly more reputable, news outlets; the development of an attention economy characterized by information overload and a reduced opportunity to critically

---

* Examples abound of such appearances but here are just two. Shortly after Donald Trump's victory in the 2016 Presidential election, in an interview on the *Diane Rehm Show* on National Public Radio, prominent Trump surrogate and CNN political commentator Scottie Nell Hughes pronounced: 'There's no such thing, unfortunately, any more as facts . . . and so Mr Trump's tweet(s), among a certain crowd, a large part of the population, are truth.' It is a sentiment shared in Moscow by Dmitry Kiselyov, journalist and former TV presenter appointed by Russian President Vladimir Putin to head up the state media conglomerate *Rossiya Segodnya*. 'Objectivity is a myth,' Kiselyov once infamously declared, 'that is proposed and imposed upon us.'

evaluate content; indwelling categorical biases in neuro-marketing, micro-targeting, search engine and social media algorithms which distribute and circulate such content not on the basis of veracity but on user demand and preference; and the proliferation, across all forms of news media, of scandal, propaganda, hoaxes and plagiarism combined with a general shift in editorial ethos towards sensational-ism, tabloidization, soft news and infotainment. All these factors, or an amalgamation thereof, have been identified over the past few years as comprising the cultural and societal harbingers of a denialist, post-truth era.

And with good reason. While, on the one hand, it's certainly true to say that even with the best will in the world most of us will seek justification for discrediting what we choose in order to engin-eer a pretext for doing as we please – anti-vaxxers, climate-change deniers and 9/11 conspiracy theorists take note – it's equally true on the other that clicking, swiping, tapping and scrolling have made it easier than ever before not just to gloss over the facts but to paint over them completely. To pretend that they don't exist. Not now. Not then. Not ever.

But I think there's something else, another contributing factor in the rising tide of denialism, fake news and the upsurge in sub-jective objectivity that speaks more to questions of identity than it does to those of online obfuscation or an appetite for the glib and sensational. It's a deeper consideration that brings us right the way back to what Piers Morgan was saying about groups and to the principle of accentuation. To how the presence of a line or a border doesn't just create differences. It exaggerates them.

Consider, as we saw in Chapter 10, the social psychologist Henri Tajfel's outstanding contribution to the advancement of human knowledge: how to get a bunch of people who've never met before to dislike each other. The solution, he discovered, was monumentally simple. You divide them into two arbitrary groups (the red group ver-sus the blue group; the warriors versus the hawks; under-estimators versus over-estimators . . . it doesn't matter what you call them) and give each group time to gel. In no time at all individuals will start to show favouritism towards members of their own group and antagonism towards those of the other. So deeply hardwired is our need for affiliation that even these so-called 'minimal groups' have the power to make us loyal.

These days, it goes without saying, with the proliferation of hashtags, online forums and social networking sites, minimal groups have a habit of springing up all over the place and our highly responsive group-loyalty buttons are constantly being pressed. Often, the psychological glue holding such groups together is a dislike of, or an antipathy towards, a specific cause, individual or ideology. But even in such instances when this is not the case, one thing is beyond question. More groups coexist *now* than at any other point in our history. Which means that group *identity* has never been more salient or important, and never more 'squeezed' or perceived to be under threat.

'Social identity lies in difference,' observed the French philosopher and anthropologist Pierre Bourdieu in his 1979 opus, *Distinction*, 'and difference is asserted against what is closest, which represents the greatest threat.'

That difference, that sense of identity, can kick in anywhere. And there's nothing like a dividing line to trigger it.

Take, as just one example, the online music-streaming service Spotify. At the time of writing, Spotify currently lists over 4,000 online categories and sub-categories of music ranging from Nintendocore – a genre delineated solely by the overlay of bludgeoning, bowel-stewing grunge riffs on old-school videogame soundtracks – to Pornogrind, a maximum security subgenre of Grindcore (a convulsing fusion of raw-boned, buzz-sawing death metal and tightly coiled hardcore punk).

You get the picture. We humans run on difference. Though a computer might struggle to systematize and codify a meaningful distinction between two, to all intents and purposes, essentially interchangeable forms of music, the rest of us seem intent on divining one. Case in point: Vegan Straight Edge versus Hatecore, a pair of musically monozygotic genres which would appear, to the uninitiated, to share a hundred per cent of their sonic DNA. Yet don't. If you are wholly unfamiliar with the categories you'd have trouble telling them apart. But like the parents of identical twins the two opponent groups of diehard aficionados would be only too happy to expatiate at length on their key distinguishing features. That festering percussive dimple. That putrid lyrical mole.

It's a fundamental problem that we, as social creatures, are encountering increasingly often as the gravitational pull of the internet draws us ever more tightly together: over-compartmentalization of the identity space and the attendant 'identity anxiety' that will inevitably accompany such categorical overcrowding. We divide not so much to conquer. But to concur. We label our beds and then lie in them – with consenting, likeminded bedfellows seeking similar categorical berths.

And the more nuanced and gradated this interpersonal branding – the finer the lines, the shorter the distance, the closer the proximity between us on the tight, competitive shelf-space of collective self-identity – the more heightened the sense of identity that we'll have and the greater will be our desire and resolve to maintain it. Rather than adapt to the mores and norms of society, even if quite clearly we're in a distinctly small minority, we resort to doing the opposite. *We* entreat *society* to fall into step with us.

In October 2018, students at the University of Manchester voted to veto applause and to replace it with the British Sign Language equivalent of 'silent jazz hands' in the interests of demonstrating greater inclusivity to those with autism, sensory issues or deafness. A year later, their counterparts at the University of Oxford followed suit. In November 2019 the actors' union Equity issued new guidelines to phase out the term 'Ladies and Gentlemen' in UK theatres in favour of gender-neutral terminology in a pledge to become more inclusive to trans members of both casts and audiences respectively. A month later, in December, easyJet bosses did the same on their flights after the airline was blasted as transphobic by a user on Twitter. 'Dear @easyjet, are you in some kind of competition to see how many times you can reinforce gender binaries?' wrote Dr Andi Fugard, a trans lecturer in social science research methods at Birkbeck College, University of London. ' "Ladies and gentlemen, boys and girls", perfume strictly segregated again by "ladies and gentlemen". Ditch sir/madam too. An organization as huge as yours must do better.' Within 24 hours, easyJet had taken Dr Fugard's suggestions on board and have since issued guidance to both pilots and cabin crew on how to be more 'inclusive and welcoming'.

Identity is in demand. More so now than at any other period in history. It must be defended and fought for as such.

To bring into sharper focus this complex, conflicted relationship between identity squeeze and identity self-defence, let's return to the theme of music. Imagine you're a note on a musical instrument, let's say a guitar, or a violin, or a harp: A, B, C, D, E, F or G. Think back, for a moment, to that time in your evolutionary history when first you were plucked (or strummed or picked or bowed) from the infinitely continuous sound space of tonal possibility and were afforded your acoustic label, let's say D for the sake of argument. You probably felt quite comfortable in your unique, heptatonic skin. You had C on one side of you and E on the other but there was sufficient sonic latitude between you, enough harmonic wriggle room, for you to feel relatively secure in your discrete, incremental identity. You didn't have to fight to be heard. People knew what a C sounded like. And they knew what an E sounded like. And both sounded different to you.

But then came the sharps and the flats. The semitones. And with them diatonic diversity. C sharp slotted in between you and C on the one hand while E flat set up shop between you and E on the other. Suddenly you start to get twitchy. These new kids on the block are a little too close for comfort, you muse. They're trespassing on my turf. Phonic anxiety begins to kick in. Do I still stand out as much as I would like? Do people still instantly recognize me as soon as I am played?

'I'm going to have to concentrate much harder than I used to on being a D,' you think to yourself. 'Now that these blow-ins have arrived on the scene I need to tread my particular frequency more carefully.'

Then . . . disaster! You hear on the grapevine that there is talk of the emergence of quarter tones. Rumour has it that there is a strange new tribe of exotic, indigenous notes trailing around the Levant and the medinas of the Arabian Peninsula that identify neither as sharps, or flats, or majors like yourself but somewhere in between. Now you get *really* worried. What happens if these transharmonic immigrants start making their way over here, you wonder. My heritage will be destroyed. As will that of my six fellow majors. All this talk of there being room for everyone in a musical scale is just plain nonsense. There isn't. And we're full up as it is. No, this has gone far enough. Us majors will have to club together and make a stand. What the sharps and the flats do is entirely up to them but if they're smart they'll follow suit. Because it was them, don't forget, who started all this . . .

Strike any chords?

In recent times similar notes have been sounding in the news. Notes of chaotic, cacophonous categorization. Lines are becoming fuzzier. Boundaries blurrier. Room in the personal identity 'scale' tighter and more constrained. As we stand at the moment there are, as we've discovered, seventy-plus gender categories on Facebook. Forty-plus different options for sexual and romantic orientation. That's not embracing the greyscale of reality. It's dicing *optimal* black and white into mini, micro, *sub-optimal* black and white.

Last year, Anthony Ekundayo Lennon, a 54-year-old white theatre director of Irish ancestry, scooped a sizeable chunk of a

£400,000 Arts Council fund aimed at supporting actors from ethnic minorities following a lifetime of mistaken identity. (Lennon's dark skin, curly 'afro' hair and distinctive facial features give him the appearance, if not of being black, then certainly of being mixed race.) 'Everybody on the planet is African,' articulated Lennon. 'It's your choice as to whether you accept it. Some people call themselves a born-again Christian. Some people call me a born-again African. I prefer to call myself an African born again . . . although I'm white, with white parents, I have gone through the struggles of a black man, a black actor.'

And he's not alone in this way of thinking. The year before last, a sixty-nine-year-old pensioner in the Netherlands who self-identifies as twenty years younger after doctors told him he has the body of a forty-five-year-old sought to legally change his age to that indicated on his biological passport so that he could return to work and enjoy more success on the dating app Tinder.

'[Transgender people] can now have their gender changed on their birth certificate,' he argues, 'and in the same spirit there should be room for an age change.'

Whatever the rights and wrongs of these issues, and it really doesn't matter which side you're on, one thing is for certain. In the current social climate of global swarming where rapidly proliferating age, gender, ethnicity and sexual orientation channels are flooding the identity airwaves and competing for space with the traditional, binary stations of male and female, black and white, gay and straight, and young, middle-aged and old, the border points and boundary posts that since the days of our earliest ancestors have shored up the differences between us are becoming finer, fuzzier and flimsier by the day. Identity bandwidth is shortening. And the prospect of adjacent-channel interference between the fragile and fluctuating frequencies of character, personality and selfhood is now greater than ever before.

Which means that we're more protective of our dials; that we've become highly resistant to even the slightest tweak or twiddle left or right and super-attuned to maintaining quality signal strength. Which means, in turn, that when like sheepish shoppers faced with the dilemma of an unrefundable coat we have the option of either holding our hands up and renouncing our beliefs ('What the hell was I thinking when I bought this?') or constructing a truth to align our beliefs with the facts ('You know what? It's actually not that bad'), we do neither. Admitting we're wrong has never been one of our strong points. And now truth has a problem, too. Even the most minuscule of nudges this side or that side of our core, fundamental belief structures, just the slightest deviation from our unique, individual flight paths, and we're in danger of entering another identity airspace. Of losing our bearings and quite possibly, if the incumbent of that airspace is hostile and minded to retaliate, being buzzed and shot out of the sky.

So what do we do? We do the only thing we can do under the circumstances. We lie. We deny. We pretend. To ourselves. To others. To the world. We 'ideologize' information. Not all of us. But a new, established and increasingly vocal minority. Might your core beliefs about climate change be threatened by a damning intergovernmental report on temperature anomalies, CO_2 emissions and rising sea levels? Might the 'identity switch cost' of changing your mind or adjusting your position on global warming and greenhouse gases be too great to take it on board? Not a problem. Denounce the report and the data contained within it as hot air.

Stricter gun control laws not to your liking? No big deal. Repudiate media reports of mass school shootings as fake news. Or, at the very least, as orchestrated by the government as part of an insidious and elaborate plot to undermine the Second Amendment.[*]

---

[*] It's important to bear in mind, of course, that conspiracy theories aren't new. At the time of writing, Coronavirus is wreaking havoc around the world and there have been mutterings from certain quarters that COVID-19 was originally

This, as we've seen, is what happened in the case of the Marjory Stoneman Douglas High School shooting in Florida. And it's a reaction that's not without form. Led, among others, by the far right conspiracy theorist and US radio host Alex Jones, the same thing happened after the Sandy Hook Elementary School shooting in Newtown, Connecticut, back in 2012 which left twenty-six – twenty students and six staff members – dead.

We live in an age of unprecedented interconnectedness, unparalleled self-exposure and unlimited public reach where belief, opinion and personal ideology comprise a greater part of our social and cognitive identities than they ever have before. Perspicacity comes veiled in persona. Integrity in image. So when faced with a challenge to the way we think and act we defend not so much ourselves but our selves. We demean. We disparage. We discredit. We

---

colonized in a Chinese military laboratory before somehow managing to escape. While pointing the finger of blame at a denigrated out-group may well serve as a tried and tested means of stoking up nationalistic fervour, facilitating in-group cohesion and thus consolidating in-group identity during periods of fear and uncertainty, it doesn't take a psychologist to flag up the benefits of unsubstantiated rumour-mongering and paranoid accusation. In the fourteenth century, when the Black Death was charting its deadly course across Europe, people were similarly incognizant of the origin of the illness. Soon afterwards, word began to spread that the Jews were responsible for the pandemic by poisoning wells in a bid to control the world. Pogroms and displacement duly followed. By the same token, in an attempt to explain the source of the so-called Spanish flu outbreak that is thought to have killed up to 50 million people between 1918 and 1920 – some 30 million more than World War I which concluded around the same time that the virus first appeared – many eyes turned towards the Germans. At Ypres, in 1915, in one of the first recorded uses of chemical weapons on the battlefield, German forces had released chlorine gas from pressurized cylinders. Might they be at it again? Might the pathogen have been secreted into aspirin manufactured by the German pharmaceutical firm Bayer? Might German U-boats have stolen into Boston harbour jingling with toxic vials and let loose the germs in theatres, railway stations and other crowded places in an act of bioterror? Such lurid attempts to make sense of the unfolding crisis encapsulated the mood of panic, dread and suspicion at the time.

think not like scientists trying to solve a puzzle but like Rottweiler lawyers trying to win a case.

If we can't turn back the clock and get a refund on our beliefs, if we can't construct a truth to get fact and fiction singing from the same hymn sheet, we choose the only option available to us to help us restore a healthy sense of internal cognitive consistency, to preserve a balanced and coherent relationship between what we think and believe on the one hand and what we say and do on the other.

We change the facts of the situation themselves. Whatever those facts. Whatever the situation. We dismiss. We falsify. We accuse. We stake fraudulent claims on little plots of publicly owned reality. We hammer up the fences and the 'Keep Out' notices. And turn them into private property.

Get real

I am sitting with Dominic Abrams, the psychological mastermind behind the optical illusion study. A long time ago, when I was a student, Dominic was my mentor and we've stayed in touch over the years. Today we're ensconced in a café near the British Academy in London off The Mall, where he is vice president of the Division of Social Sciences. I tell him my theory of modified cognitive dissonance, of the slaughtering of fact on altars of precision identity, and he authors enough sage nods and strokes his chin a sufficient number of times to suggest I may well be on to something.

The problem is, what to do about it? If reality continues to fragment into a balkanized battlespace of nervy, totalitarian micro-identities then eventually nothing will be real and everything will be real. We'll wind up back in 1984 with George Orwell, only for completely different reasons. Devolved, decentralized reality is just as bad as its authoritarian, state-sponsored counterpart. For reality to fly it needs to be democratic.

'At the heart of the issue lies a giant paradox,' Dominic explains.

'Bigger than anything the ancient Greeks ever came up with. We are inherently social creatures whose inherently social nature compels us to behave in inherently *antisocial* ways. To stem the tide of the post-truth revolution, to arrest the development of disinformation, fake news or whatever you want to call it, we need to think about why people join groups in the first place; the reasons that we have – that we have always had – for worshipping this god or that god, supporting this team or that team, voting Republican or Democrat, Conservative or Liberal.

'They haven't changed. They've been the same for centuries, millennia, right the way back to prehistory. Yes, we need security and status. Yes, we need to belong. But we also need to be right. Or, at the very least, to convince ourselves that we're right. To have enough people around us thinking and feeling the same way to give us the impression that our perception of the world, our attitudes and actions, our beliefs and behaviour, are accurate, virtuous and appropriate responses to the situation, whatever situation that may be.

'So, if we can somehow find a way of meeting these needs but do so in a way that doesn't trigger partisan loyalties, that doesn't create tribal tensions, that doesn't, in short, cultivate an us–them mentality, then it'll be a step in the right direction.'

It certainly would. The trouble, of course, as I respectfully point out, is that no one has actually managed to achieve this yet. No one, at any time in history – ancient, modern or pre – has succeeded in pulling it off. Because us–them is as old as the hills, as strong as the sea, and as set in its evolutionary course as the stars across the heavens. Taking the us–them out of us and them would be like taking the hot out of fire or the wet out of water. It simply can't be done.

Moreover, would we even *want* to do it? On the one hand we might aspire to rid ourselves of the zealots: the biased, the bigots and the firebrand fundamentalists. But consider what else we'd be losing in the process. Super Bowl. The Beatles. Brexit. The Eurovision Song Contest, for heaven's sake.

We sculpt who we are by chipping away who we're not.

Where do we draw the line between those who present with a full-blown Group Identity Disorder* and those who are 'acceptably biased', high on the spectrum but short of the clinically partisan? Just as there exists a slender stock of basic and primary hues within the infinitely continuous colour space and a skeleton staff of just seven primary notes in a standard musical scale, where, as a society, do we set our identity viewfinders to give us the optimal number of channels on the innumerable airwaves of self? How do we attain that elusive diversity sweet spot with the maximum number of tribes yet the minimum amount of confusion, the lowest degree of conflict and the least categorical disruption to our powers of communication?

It's an impossible question to answer. And as for who should answer it, who should be tasked with calibrating those viewfinder settings to determine, within any particular core demographic network – gender, political affiliation, sexual orientation, race – the appropriate channels and frequencies, their strength and number, and the degree of separation between them, that's an issue in itself. The government? The courts? Policy institutes? If Orwell were with us today and working on 2084, would we have not just the thought police but the category police as well?

In the UK, during the Coronavirus crisis of 2020, we virtually reached that point with one of the starkest viewfinder dilemmas that any generation is ever likely to face. Do we zoom in and think of ourselves as individuals: go out, visit friends, go to work? Or do we look at the bigger picture and consider the old and the vulnerable? The government, in the end, answered the question for us by

---

* As yet a fictitious condition, but clinical taxonomists take note.

introducing strict guidelines on social distancing (everyone should keep at least two metres apart), employment (who could continue working and who couldn't), and supermarket etiquette (imposing limits on the amount of shopping we could buy). Parliament, in other words, nationalized our personal viewfinders and set them all firmly on landscape. The measures proved broadly effective. With 'them' on mute, and militant self-interest faded out, there across the existential airwaves just the one, single, unwavering note of identity.

'Us'.*

But whether those measures appealed to our better natures or, as *Flipnosis* might predict, inadvertently fired the influence silver bullet of self-interest, is a moot point. In the early stages of the crisis, just as the virus was beginning to take hold, Jim Everett, one of Dominic's colleagues at the University of Kent, conducted a study to see which type of message might be the most effective in getting people to comply with the guidelines laid down by the government. The answer, it turned out, were those that emphasized the role of everyday duty and responsibility in keeping friends and family safe, as opposed to messages which focused more explicitly on the moral aspects of embracing the 'new normal', or, relatedly, on the utilitarian perspective of the health consequences for everyone if we didn't.

But what was conspicuous by its absence from the study was another kind of message: one which reflected the social cost to the

---

* Note, here, the role played by context in determining the correct viewfinder setting. Cast your mind back to the previous chapter and to the heroic British captain who lost his life returning a wounded German soldier to his compatriots on the front line, and his viewfinder, you may recall, was set on close-up. When we're not on the same side there is virtue in focusing on individuals as opposed to groups. When we *are* on the same side, however, as we were in the fight against Coronavirus, then the utility of the settings flips. Zooming out and focusing at a group as opposed to an individual level constitutes the overriding moral imperative.

individual of disregarding the new measures – shaming, disapproval, ostracization.* This, I suspect, may well have been the most powerful message of all and could quite easily have been the thinking behind the nationwide text sent out by the UK government to all mobile phone users informing recipients of the tightened restrictions on leaving their homes. With everyone in the loop there could be no complaints or excuses.† The more of us are 'us' the greater the price on 'them'.

Dominic picks up the thread.

'Of course, I'm not trying to write a John Lennon song here,' he's at pains to point out. 'I'm not saying, "Imagine a world with no us and them, just us." What I *am* saying, though, is that there are

---

* An April 2020 study examining how the content of political rhetoric has changed over the years reveals that language related to in-groups and togetherness (most commonly, terms such as 'nation', 'community' and 'unite') began to usurp language relating to duties and obligations (most commonly, 'law', 'order' and 'authority') at some point during the Great Depression of the early twentieth century. This significant shift in the narrative of political persuasion from that emphasizing rule-following to discourse relating to unity and group identity is predicated almost entirely on three factors: the rise of Western nationalism during this period, the emergence of a truly pan-national consciousness, and the mass movement of people from agrarian to urban societies. The study in question, led by Nicholas Buttrick, a social psychology graduate student at the University of Virginia, analysed over 7 million words of political speech – 1,666 documents in total – from three different countries (America, Canada and New Zealand), the earliest of which dated right the way back to 1789.
† It should be noted that police in Derbyshire were criticized for being 'heavy-handed' following the introduction of the social-distancing measures after they used drones to photograph people out walking in the Peak District and then posted the photographs on social media – a decision which subsequently led to calls for consistency across police forces in implementing the new guidelines. To this end, the government, in partnership with the Police Federation of England and Wales, the National Police Chiefs Council, and the College of Policing, formulated a common-sense approach – as defined by the 'Four Es' – to ensuring that the new regulations were followed: Engage (encourage voluntary compliance); Explain (point out the risks of non-compliance); Encourage (emphasize the benefits of compliance); and Enforce (either direct – or remove – people back to the place where they live using reasonable force where necessary).

certain times, certain contexts and situations, when it's important for us to keep a close eye on just how we *manage* our need for belongingness and all that it entails.

'It's not difficult. The science is there. All we need is the will. We know, for example, that creating an environment of psychological safety in which the risk of compromise to an individual's worldview or self-esteem is mitigated by an emphasis on having an open mind, of being curious, inquisitive and not feeling threatened by critical feedback or disconfirmatory evidence, facilitates innovation and the quest for novel solutions – and that taps into our fundamental need for identity. For autonomy, self-determination and the need, I guess, to be right.

'Alternatively, or simultaneously even, incentivizing the obverse need not to screw up by promoting a culture of accountability in which individuals are held responsible for their mistakes, for erroneous actions and incorrect decision-making, achieves much the same end.'

Research also indicates, Dominic continues, that whereas we generally tend to push back on influence from out-group sources to preserve our in-group identities, messages and interventions that appeal to the notion of a superordinate identity, such as Americans, Europeans or ultimately – one thinks of the ongoing conversations surrounding climate change and, as we saw in 2020, the fight against COVID-19, for instance, and the death of George Floyd* – human beings, can dramatically reduce group-related prejudice and bring all of us together, from the smallest, remotest islands to the biggest brashest cities, from the richest, most powerful billionaires to the single parent in and out of work, under the one persuasion roof.

Midway through 2019, the UK population was split down the middle over Brexit. Battalions of Leavers massed on one side. Squadrons of Remainers on the other. So sharply were the battle

---

* As a Twitter meme at the time put it: 'It's not black versus white. It's everyone against racists.'

lines drawn that relationships had ended on the strength of them and families were bitterly divided. Six months later, in the March of 2020, we were never more united.

Coronavirus may have claimed the lives of individuals. Every single one of them a tragedy. But as a nation it might've saved us.

In one study, Dominic tells me, two groups of participants comprising four members each (AAAA and BBBB) convened in separate rooms to discuss the solution to a problem (the so-called 'Winter Survival Problem': their plane had crash-landed in the woods in the middle of January and they had to rank order items salvaged from the plane – a gun, a newspaper and a tub of lard, for example – in terms of their importance for survival). Both groups then came together as one around a conveniently octagonal table to hammer out a joint proposal. But there was, as always, a catch. This time with the seating plan. In one variation of the study, the groups remained fully segregated (AAAABBBB). In another they were partially segregated (AABABBAB), while in a third they were fully integrated (ABABABAB).

The effects were incredible. Not only did this rudimentary game of psychological musical chairs significantly reduce in-group bias among those who were fully integrated, it also increased cooperation levels, friendliness ratings and respective member confidence in the jointly proposed solution.

In another study, he goes on, conducted in the States at the University of Delaware football stadium, a mixed pool of black and white interviewers – some wearing a home-team hat (rendering them a member of the common, salient in-group) and some an away-team (Westchester State University) hat that conferred no such extenuating affinity or superordinate common ground – canvassed the opinions of *white* home-team fans on the kinds of food they liked.

Of most interest to the researchers, needless to say, was the performance of the black interviewers. With which of them would the fans engage more? The answer, obvious when it's right in front of you but not so obvious that practitioners and policymakers don't pay more attention to the power of the superordinate in blindsiding difference and prejudice, is exactly as you'd expect. Those who were wearing the home-team hats.

'And the reason for that is simple,' Dominic explains. 'In the niche partisan atmosphere of the football stadium, the cultural behemoth of race suddenly got bumped down the context-specific pecking order of group identity by the imperative of sporting allegiance. Momentarily, who you supported *now* became a lot more important than where you came from *then*.'

'So team colours won out over skin colour?' I suggest.

Dominic smiles.

'Indeed they did,' he says. 'Which, of course, you would've known had you actually turned up for some of my lectures. Bet you wish you'd paid more attention all those years ago now, don't you?'

He's right. I do wish I'd paid more attention. Except in those days 'us' and 'them' were just that little bit thinner on the ground than they are today. They still existed, of course. They were still doing the rounds. You were either 'Blur' or 'Oasis' as far as I recall. And there were five in the Spice Girls on which to pin your identity. In fact, if Spotify were to welcome another musical label to the fold and were to call it 'identity pop' then Posh, Sporty, Scary, Baby and Ginger would have a case for founder membership. But there certainly weren't seventy-odd different genders on Facebook. Because, of course, there wasn't Facebook either.

Regardless of rehab and reason, enshrined in paradox and the endogenous absurdity of self, the truth that emerged a long, long time ago by the rivers and lakes, within the canopies and caves, around the kill sites and campfires of old, old Africa has, like us, stuck around. And is, like us – wars, pandemics and carbon emissions apart – here to stay.

Without a 'them' there can never be an 'us'.

It's the others that bequeath us ourselves.

The Wisdom of Radicals

∴

Colour is everything. Black and white is more.

DOMINIC ROUSE

O N 29 JUNE 2016, forty-one-year-old Emirati national Ahmed Al Menhali, owner of a business marketing company in Abu Dhabi, was standing outside a hotel in Avon, Ohio, talking in Arabic on his mobile phone when he was accosted by a bunch of armed police officers yelling at him to get on the ground. Three weeks later, in Palm Coast, Florida, two teenagers sitting in a car playing the augmented reality game Pokémon Go find themselves being shot at by a local homeowner armed with a handgun. Then, on 23 February 2017, at Levenshulme train station in Manchester, police Taser a blind man with a 50,000-volt stun gun.

Three isolated incidents, but all have something in common. In each one the victim in question suddenly finds himself at the wrong end of a malign categorization error that subsequently leads to a grotesque distortion of reality on the part of his assailant.

In the case of the Emirati businessman, the observation that he was wearing a traditional white *kandura*, or ankle-length robe, and headscarf, and was talking in his native tongue prompted the sister of a hotel clerk to dial 911 reporting 'a suspicious man with disposable phones, two of them, in a full head-dress . . . pledging his allegiance or something to Isis'.

In the case of the Florida Pokémon Go players, the fact that the homeowner was awoken in the early hours by 'a loud noise coming from outside his home'; the fact that he had, on further investigation, come upon two teenagers sitting in a car; and the fact that he had, as he approached the vehicle, heard one of them say something like 'Did you get anything?' led him to believe they were criminals. Subsequently, when the man stepped in front of the car, ordered the teenagers not to move, and it accelerated towards him, he interpreted the advance not as an act of fear but one at best of evasion, and at worst of attempted homicide.

Finally, in the case of the blind man, it was his foldaway walking stick that had proved the offending instrument, triggering a call to police that there was a man with a firearm wandering about in the train station. By the time they arrived the officers in question had had plenty of time to pigeonhole the man in the erroneous category 'gunman', just as the Ohio police had had plenty of time to stereotype the Arab businessman as a member of the category 'terrorist'. And so when they turned up on the platform, Tasers and attack dogs in tow, rather than 'seeing' a blind man with a folded-up cane innocently waiting for his train home, they saw a lunatic with a handgun instead.

Psychologists and brain scientists have been on the tail of this dark side of categorization for quite some time. And with good reason. Though no one actually lost their lives in the aforementioned incidents, in all three cases the psychological trauma incurred by the victims was considerable. Even so, they can count themselves as lucky. Sometimes, those on the receiving end of categorization errors are not so fortunate. End up in the wrong place at the wrong time and the consequences can be catastrophic.

Shortly after noon on 9 August 2014, on Canfield Drive in Ferguson, Missouri – a northern suburb of St Louis – Darren Wilson, a twenty-eight-year-old white law-enforcement officer, shot eighteen-year-old Michael Brown six times as the black teenager ran towards him with his right hand shoved ambiguously into his waistband under his shirt. Moments earlier there had been an altercation between Wilson and Brown during which Brown, a suspect in a theft at a local convenience store, had repeatedly landed punches on Wilson through the window of his patrol car, had allegedly made a grab for his gun, and had then run off when the weapon, accidentally or otherwise, discharged during the ensuing struggle. Getting out of the car Wilson subsequently set off in pursuit of Brown, took a shot at him and missed, while continuing to yell at him to stop and get on the ground.

Eventually Brown did stop. But, rather than heeding the lawman's instructions and prostrating himself, he emitted, according to Wilson, a 'grunting noise' and began running back towards him. It was Wilson's final bullet, shattering the top of Brown's skull, that ended the young man's life.

The shooting of Michael Brown made headlines around the world. Plaintive memorials and quivering candlelit vigils soon gave way to full-scale civil unrest as members of Ferguson's predominantly black community repeatedly clashed with law-enforcement officers – armed, post-9/11, with military-grade weapons and bedecked in riot gear – over their strong-arm annexation of several sensitive areas of the city. In the immediate aftermath of the killing, eyewitnesses claimed that Brown had been shot in the back while running away. Others that he'd had his hands up. Race was at the heart of the issue.

In due course, the tortuous legal process finally at a close, the grand jury made an announcement. Darren Wilson would not be indicted for the murder of Michael Brown. Tensions erupted again. There was, concluded Robert P. McCulloch, the prosecuting attorney for St Louis County, insufficient evidence to disprove Wilson's assertion that he discharged his weapon in fear of his own safety, and moreover, both forensic evidence and reliable witness statements corroborated Wilson's report of the events that had unfolded on Canfield Drive that fateful afternoon.

'We may never know exactly what happened,' commented the then US President, Barack Obama. 'But Officer Wilson like anybody else who is charged with a crime benefits from due process and a reasonable-doubt standard.'

Debate still rages over whether justice was done in Ferguson. In November 2014 Darren Wilson resigned his commission in the town over security fears. He hasn't worked since. In May 2016 Lezley McSpadden, Michael Brown's mother, brought out a book to tell her son's side of the story. Of video footage showing Brown handing over a suspicious package in the convenience store he was suspected of robbing just moments before he was shot, and then aggressively shoving an attendant who appears to remonstrate with him as he makes his way out, she is, understandably, circumspect: 'Eighteen seconds . . .' she writes poignantly, '. . . doesn't tell you anything about eighteen years.'

Obama was right. We'll never know with absolute certainty the true nature of events as they unfolded that day in Ferguson. But what we can be certain of is this. Subjective categorization of intent, motive and behaviour on the basis of race can, quite literally, make people see things that aren't there, whether it's a gun, a lunge for a gun, or hands raised in surrender.

Keith Payne, professor of psychology and neuroscience at the University of North Carolina, knows this better than anyone. Back in 2001, some thirteen years before Darren Wilson and Michael Brown first clapped eyes on each other, he set out to answer one simple question: does the way we categorize race really influence what we *see* or are the effects of categorization more subtle and 'psychological' in nature?

To find out, he conducted a study in which he showed participants – the usual cohort of college students – a picture of either a

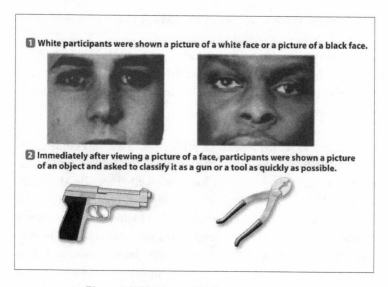

Figure 1: White versus black; guns versus tools.

white face or a black face followed immediately afterwards by an image of either a hand*gun* or a hand *tool* (e.g. pliers, a socket wrench or an electric drill – see Figure 1 above). The task was simple: to identify the object – gun or tool – as quickly as possible.

What he discovered was a wake-up call to all of us. In an eerie, empirical premonition of the events that were to unfold a couple of states west some decade and a half later, not only were the students – hardly at the top of most people's lists of racist, xeno-phobic bigots – better at identifying *weapons* after they saw a black face, and better at identifying *tools* after they saw a white face, they also, as Payne reports, 'falsely claimed to see a gun more often when the face was black than when it was white'.

But why? What is it, precisely, about the category 'skin colour' that gives rise to such knee-jerk misjudgements?

A 2006 study by a team of neuroscientists at the University of

Colorado in Boulder, led by associate professor of psychology Josh Correll, provides the answer. Correll and his colleagues had participants play a video game, known as the Weapons Identification Task, in which they were presented with short, sharp, rapid-fire images of armed and unarmed men – some black, some white, some brandishing a gun, some holding a wallet or mobile phone – and had to make split-second decisions about whether or not to shoot (see Figure 2 below).

Figure 2: Armed and unarmed images: Left – European American holding a wallet. Right – African American holding a gun.

The researchers were interested in whether race had a say in participants' reaction times on the task; whether the perceptual biases reported by Payne five years earlier actually translated into an overtly aggressive response pattern. But that wasn't all they were interested in. They also wanted to know what was going on inside participants' brains as they grappled with the different contingencies, with the fleeting, flashing life-or-death decisions that peppered the screens in front of them.

To do that, they wired each of them up to EEG apparatus while they played the game and measured something called event-related brain potentials (ERPs) – fluctuations in the electrical activity of the brain that are a direct result of exposure to a particular stimulus.

If Payne's findings weren't provocative enough, then those of Correll and his team took things to a different level. Not only were participants quicker on the draw when confronted with armed black as opposed to armed white suspects, they were also quicker to decide not to shoot unarmed whites than unarmed blacks.

But that wasn't all. The results of the EEG revealed why. Probing deep inside participants' brains as they performed the task, the researchers discovered that far from processing the various scenarios using an objectively binary, dispassionately dualist method of categorization – armed versus unarmed – they were in fact employing a taxonomic framework that was, though binary, anything but objective and dispassionate.

Black versus white.

To be more specific, participants were quick to pull the trigger when confronted with armed whites and armed blacks. No problem there. Trouble was, they were also trigger-happy for *unarmed* blacks as well. The only stimulus category that didn't set alarm bells ringing was unarmed whites.

The bottom line, derived by Correll and his co-authors from evolutionary theory, seemed unavoidable. 'To survive, humans must reliably detect threats in their environment,' they concluded. 'Attentional processes, and consequently ERPs, should therefore differentiate between threatening and innocuous stimuli. In the context of the game, armed targets constitute a threat: they are the "bad guys". But, in light of cultural stereotypes, even unarmed blacks may be perceived as threats.'

Payne and Correll's research demonstrates three basic principles inherent to the process of categorization that we've encountered a number of times throughout the course of this book. First, that the practice is fundamental to survival. Secondly, that it is often unconscious. And thirdly, that it is hallucinogenic. We don't, in other words, categorize what we see. We see what we categorize. In fact, just *how* fundamental, unconscious and hallucinogenic this categorization instinct is may be gleaned from a simple study conducted by psychologists at the University of Pittsburgh back in 1980. Researchers presented a bunch of sixth-grade school kids with a series of line drawings depicting two students interacting with each other in a variety of ambiguous scenarios. The scenarios were accompanied by oral descriptions of the nature of those interactions. One boy bumping into another in the hallway, for example. Or one boy asking another he didn't know very well for his cake at lunch.

But the experiment, though simple, wasn't quite as straightforward as it seemed. The line drawings weren't all the same. There was a subtle, but crucial difference. Half of the hallway bumpers or cake hijackers portrayed in the pictures were white while the other half were black. Moreover, the sixth-graders were themselves, for the purposes of the study, split into two groups. White and African American. That said, their task was easy. They had to indicate, on a seven-point scale, how friendly or threatening they thought the ambiguous bumps and would-be cake hijackings were.

Would the ethnicity of the characters depicted in the cartoons influence the schoolkids' ratings, the researchers wanted to know? Furthermore, might this also depend on the ethnicity of the kids themselves?

In the first case the answer was yes. In the second, no. Results showed that when the bumpers and hijackers were black their actions were perceived as significantly meaner and more threatening than when they were white. But incredibly, this assessment bias held firm irrespective of the ethnicity of the rater. Whether the school kids were white or African American made little difference. All of them evaluated the conduct and demeanour of the black actors as appreciably more hostile and aggressive. Not only does categorization start early, then, but the roots of stereotyping go so deep as to trigger self-harm. To inveigle groups into exhibiting prejudice that discriminates against themselves.

But there's another side to this story. A side seldom told yet equally fundamental. Stereotyping keeps us alive.

Several years ago a friend and I were running a study in the wilds of Africa. Late one afternoon, as we're strolling along a dirt track in the middle of nowhere, he suddenly jumps ten feet in the air and lands on the other side of the road. When he eventually regains his composure he points to the side of the track and there, coiling out of the undergrowth, is a gnarled, twisted tree branch that looks exactly like a snake. Fortunately for him it wasn't and the worst he suffered was a bruised ego. But better to look silly than dead. Better to err on the side of a false positive (mistakenly believing the branch to be a snake when it isn't) than a false negative (believing it isn't a snake when it is). Better, in other words, to stereotype all snake-shaped sticks as snakes than to stand by the side of the road trying to build up horticultural rapport.

The difference, of course, is that sticks don't have feelings. Stereotyping all sticks as snakes doesn't upset the stick population of the world and it doesn't anticipate discrimination or arboreal oppression. Ride the tube, however, and stereotype a Muslim gentleman in traditional dress with a beard, a copy of the Qu'ran, and a rucksack on his back as a terrorist and you soon run into trouble. As the police did in Avon, Ohio.

But why? Isn't it the same thing? Doesn't that vague, uncomfortable notion you may have to get off the train and take the next one come from exactly the same place as jumping ten feet in the air when you mistake a harmless tree branch at the side of the road for a snake? Or drawing a gun on someone you think (rightly or wrongly) is going to shoot you?

Better a false positive than a false negative.

The truth about stereotyping is that it allows us to make quick judgments about people. Sometimes those judgments may be wrong. Sometimes they may be right. But the point is they are *quick* and sometimes quick judgments are precisely what's needed in life.

The writer Kurt Vonnegut once said something profound: 'Life happens too fast for you ever to think about it. If you could just persuade people of this, but they insist on amassing information.' He's right. There's a lot written about the negative side of stereotyping and there's no denying that stereotyping can be a problem. But equally there are times when it can also save our lives. That's why our capacity to make big decisions based on small data – to favour efficiency over accuracy – evolved in the first place.

Take another example. Imagine you find yourself in a strange city at night walking along the pavement and four teenagers in hoodies appear out of nowhere and start walking behind you. Would it be unreasonable for you to cross the street . . . you know, just in case? Now ask yourself the same question if it was four men in business suits?

Most people give very different answers to these two questions. Stereotyping is like goalkeeping in soccer. Play well and the truth is most of the time no one will pay much attention. Unlike the strikers up front, goalkeepers rarely get the credit they deserve. But make an error and let in a goal, let the ball slip through your fingers or between your legs, and everyone will be on your case.

Stereotyping *itself* has been stereotyped. As bad rather than good. But much of the time it keeps us in the game. The trouble is we just don't notice.

We glimpse, in our ignorance, the phantoms of denial and repression. Stereotyping is an act of extremism. And extremism has got a bad rap. Cross the road when you see a bunch of hoodies coming around the corner and you firebomb the plaza of an open, inclusive mind. You blow up the benefit of the doubt. But this is the road that each of us must cross. The truth we all must face. You and me. We're extremists. You, me, Adolf and Osama. We're extremists. You and Donald. Extremists. Me and Putin. Extremists. All of us. All of us are extremists. We are born extremists. We live our lives as extremists. And we die extremists. From the moment we draw our first breath, flex our first muscles, issue our first cry, we embark on a life of extremism. We sign up to the extremist's creed. There is no backing out. No second chances. And none of us are immune. We are extremists because we *have* to be extremists. Because if we weren't extremists we wouldn't, *couldn't*, be anything.

Let's take a simple example. Let's say that you're sitting down. And then stand up. Maybe you've been working at your computer for a couple of hours and decide to grab a coffee. Whether you like it or not, that rudimentary act of standing up is an act of extremism. It may not seem like it. But it is. It may not seem the same as flying a 757 into the belly of the World Trade Center. Or blowing up a train station in Madrid. Or even crossing the street when you see a bunch of hoodies coming around the corner towards you. But at the level of basic biology, at the level of deep-dive neurology, it is identical. We either do these things. Or we don't. Entrenched within the bedrock of our brains there are common physiological roots. What vary, needless to say, are the *consequences* of these different kinds of action. What happens once we perform them. And the volume and complexity of contingent, contextual information that anticipates such performance.

Precisely *why* these actions are fundamentally the same comes down to basic biology, to the underlying mechanism by which our brains make decisions. The cells – or neurons – that comprise our central nervous systems communicate with each other via electrochemical transmission. Stimuli from the environment, both internal and external – feelings of fatigue, the alluring aroma of coffee from down the hall – elicit changes in the chemical composition of the cells which subsequently cause them to produce electrical signals.

For every decision we make, then – be it a coffee we resolve on drinking, a street we figure on crossing, or a train station we settle on bombing – our neurons take a vote on it. Firing – the production of an electrical signal – means that they're in favour of the decision. Not firing means that they're not. If the motion is carried we go for it. We do whatever it is that we've been planning, or plotting, or thinking about. If it isn't, quite simply, we don't.

But unlike voting in politics, in the House of Representatives or the House of Commons for example, abstention is not an option for our neurological electorate. A neuron either fires or it doesn't. There is no electrochemical fence for it to park its synaptic butt on. No nervy, knife-edge recounts or interminable second referenda. It is black and white. Yes or no. Take it or leave it. Our brain cells and nerve cells go about their business according to what neurophysiologists call the *all or none* principle. Which means, when it comes to what we do, that we either get up and go for a coffee or we plough on with the task in hand. What we might happen to be *thinking* about at any point in the process is irrelevant. We may well *feel* 'in two minds' about taking a break. And we might agonize about it for hours. But at the end of the day, when the ledger of what we've done and what we haven't is submitted for public audit, it couldn't be any simpler.

It's either caffeine or bust. Twin Towers or bust. The other side of the street or bust. It's either–or. There are no two ways about it.

Every decision we make involves the drawing of a line, a line between what we were doing before we made that decision and what we do after it. A line in both time ('then' and 'now') and in space ('here' and 'there'). It may only signify a difference of milliseconds or millimetres. But that is irrelevant. Action of any kind involves choice. And choice of any kind entails the selection of one path over another. Paths across time. Paths across space. We cannot be in two places at the same time. And we cannot be in two times at the same place. We are extremists from the bottom up.

Our brains are crucibles of radicalization. When a neuron makes its mind up, that's it.

Which means we have a bit of a problem. Our brains might be run by an all-or-nothing hotbed of hotheads. By a radical militia of fanatical, revolutionary neurons. But the lives we lead, the landscapes we navigate, are coded and encrypted in different categorical language, are of a different metaphysical consistency entirely. We live in a world not of binary, black-and-white dichotomies. But of continua. We inhabit an environment not of yes-or-no, this-or-that, one-or-the-other dualities. But of maybes, probabilities and grey zones.

The paradox is as stark as it is intractable. If the brain embodies an immeasurable, labyrinthine assembly of polemical, fundamentalist, do-or-die command cells, then the calls to action to which they must respond constitute alien assignments wholly incompatible with their extremist electrochemical protocols. Yet respond we do because we need to make decisions. Snakes need to be differentiated from sticks. Muggers from non-muggers. Terrorists from non-terrorists. It's just the way it is.

During the course of this book we've seen time and again how lines are our biggest protectors. The drawing of a line is the making of a difference, the creation of a category. It slaps down the threat of informational insurrection, of cognitive and perceptual disorder, by cornering the rabble of the unruly, marauding *other* – the proximal, the adjunctive, the ever-encroaching stimuli of 'the fullness of reality to be known', as the psychologist William James once put it – into segregated pens of psychologically sealed reality so that the swarming whoosh of everything is conceptually presented in singular, discrete and isolated categories, and is amenable to control and containment.

'Everything should be made as simple as possible but no simpler,' Albert Einstein once remarked.

And lines are our wands of simplicity. The problem, however, is that simplicity lies on a spectrum. And it is inestimably, inordinately complex.

The linguistics and perception of colour

Basic colour terms

In 1969, the American anthropologist Brent Berlin and linguist Paul Kay put forward the notion that differences in colour perception across cultures centre around a disparity in the number of basic colour terms that each culture possesses. A basic colour term (BCT) refers to a colour word that may be used to describe a wide variety of objects (unlike blonde), is monolexemic (unlike dark green), and is deployed on a regular basis in everyday language by most native speakers of that language (unlike magenta).

The languages of modern industrial societies incorporate thousands of colour words, but only a handful of basic colour terms. English has eleven: red, yellow, green, blue, black, white, grey, orange, brown, pink and purple. Slavic languages have twelve, with separate basic terms for light blue and dark blue.

In various indigenous languages such as those spoken by the Dani people of the Papua New Guinean Highlands or the Herero dialect native to the Himba people of northern Namibia and southern Angola, the number of BCTs can be significantly smaller, sometimes as few as two or three, with descriptive compasses that span much larger regions of colour space than the BCT descriptors of more mainstream modern languages.

In fact, when one traces the development of languages over time one discovers that the accumulation of BCTs occurs gradually. Typically, they begin as descriptors of a narrow range of objects and properties, many of them denoting physical properties other than colour such as toxicity and ripeness. These narrow descriptive parameters then progressively expand to include more general and abstract connotations before eventually assuming a pure colour sense.

Primary colours

The term 'primary colours' refers to those colours that, when combined, can be used to create any other colour. It is generally accepted that there are three primary colours – red, yellow and blue – though green is sometimes included as a fourth.

The colours of the spectrum

Spectral colours, or the colours of the rainbow, delineate those colours that together form the constituents of white light – as first demonstrated by Isaac Newton's famous deconstruction of a beam of sunlight through a prism. The general consensus is that the (visible) spectrum consists of seven colours – red, orange, yellow, green, blue, indigo and violet – though some commentators are mean to indigo and exclude it on the grounds that it doesn't work hard enough to distinguish itself from violet.

Assess your own need for cognitive closure

Decisions, decisions! We can't avoid them. Chocolate or strawberry? Caribbean or the Greek islands? Kids or career? Some of us are great at making our minds up and navigating uncertainty but others spend hours agonizing over even the most trivial things. Which camp do you fall into? Can't decide? Well, this simple quiz might help. To find out, indicate the extent with which you agree or disagree with each of the statements below. If you strongly agree give yourself 3 points, if you agree give yourself 2 points, disagree 1 point, and strongly disagree 0 points. Then add up your total and check it against the scale to find out how *your* need for cognitive closure measures up . . .

1. In a restaurant I'm usually one of the first to decide what I want to eat off the menu.
2. If I call someone and they don't answer I generally hang up in seven rings or under.
3. I'm much more of a 'plan of action' than a 'see what happens' type of person.
4. Once I've made my mind up I'm tough to persuade.
5. Waiting around for the phone to ring drives me nuts.
6. I'm very much at home with rules and routine.
7. I like things to be black and white. The more shades of grey there are the more it bothers me.

8. If I'm watching a film and have to stop halfway through I prefer someone to tell me how it ends rather than have to sit down and watch it again later.
9. Long drawn-out negotiations aren't my strong point.
10. Surprise parties freak me out.
11. I couldn't agree more with Colin Powell's 40/70 rule: that one should collect 40 per cent to 70 per cent of available facts and data when making decisions and then rely on gut instinct. Having less than 40 per cent of the facts is too risky. But continuing to gather data beyond a level of 70 per cent confidence means you might miss opportunities and others may capitalize on your hesitation.

How do you rate?

0–11: Your mind is open 24/7, 365 days of the year.

12–17: Round-the-clock thinking but closed on public holidays.

18–22: A good think–life balance.

23–28: Thinking-time regulations apply.

29–33: Alternative perspectives and other possibilities considered but strictly by appointment only.

A brief history of frames

Among the first thinkers to write about frames – or 'schemata' as they were initially referred to – was the eighteenth-century German philosopher Immanuel Kant. In his *Critique of Pure Reason* of 1781, Kant observed that we make sense of our perceptions and experiences in the here and now by comparing them with stereotypical mental representations of similar cases – schemata – stored in our imaginations. For example, we associate our present experience of the sun setting with previous experiences recalled from memory and this temporal interaction between past and present facilitates our understanding – and hence our navigation – of the social and physical environments in which we live. Schemata, in other words – or 'frames', or 'scripts', or 'scenes' – represent mental models of preconceived ideas that order, organize and manage categories of information (e.g. sunsets) and the relationships that exist between them.

Once formed, schemata (or schemas) are extremely resistant to change and exert a powerful influence over our beliefs, attitudes and judgments. For instance, the closer the fit between an object, opinion or phenomenon and our schema of that entity the more likely we are to notice, prefer and endorse it. The 'framing' of arguments or messages thus alludes to the process by which persuasive communications may be positioned in such a way as to maximize resemblance to pre-existing schemas present in the minds of their intended influence targets.

The term 'schema' was first introduced by the Swiss developmental psychologist Jean Piaget in 1923 to describe stored categories of knowledge and information present in the brains of children, the existence and modification of which inform how we perceive, learn about and interact with the world as we reason our way through the lifespan. Several years later, in 1932, the British psychologist Sir Frederic Bartlett incorporated the concept into the study of memory and learning – in particular how the presence of schemas could bias and influence recall.

It was the computer scientist Marvin Minsky at the Massachusetts Institute of Technology (MIT) who first coined the term 'frame' in the 1970s in relation to the representation of knowledge in machine learning, an approach subsequently expanded upon by the American cognitive scientist David Rumelhart, who reclaimed the idea for psychology by using it throughout the eighties as the blueprint of a neural-based algorithmic model of the mental representation of complex knowledge in humans.

The concept of 'scripts' was introduced in the seventies by two Yale psychologists, Roger Schank and Robert Abelson, and describes generic, stereotypical knowledge of action sequences and thematically linked programmes of behaviour in given situations (e.g. the script for dining in a restaurant would include a number of components including taking a seat at a table, perusing the menu, and ordering food and drinks from a waiter). One of the key mechanisms of humour lies in the disruption of such scripts through the comedic insertion of incongruous or unexpected events: one thinks, for example, in the context of dining scripts, of the famous scene from the *Fawlty Towers* episode 'The Germans' in which the madcap hotel owner Basil Fawlty (played by John Cleese) ends up goose-stepping around the tables in the hotel restaurant, reducing a party of German residents to tears.

The role of framing in persuasion and influence has been extensively studied by the Berkeley-based linguist, philosopher and cognitive scientist George Lakoff. Lakoff's specific area of interest lies in how the strategic use of language and metaphor can subtly activate stored knowledge categories and shape the way we think about particular issues. Every word, according to Lakoff, is defined in relation to a conceptual framework and, as such, has the power to trigger learned associations and acquired networks of meaning for either good or ill.

The phrase 'tax relief', for example, implies that tax is a burden, an irritation from which we require liberating. In contrast, the phrase 'pro-life' insinuates that those who support abortion are 'pro-death' and that women who seek terminations are, irrespective of their reasons for doing so, tantamount to murderers.

Berinmo versus English colour space

⁝

When colour chips of certain hues and intensities are presented to native English speakers, and observers are asked to indicate the colour of each chip respectively, the pattern of responses typically obtained mirrors that shown in Figure X below.

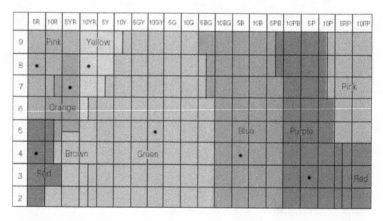

Figure X – The colour space of native English speakers (dots indicate the focal point for each colour – the point in the colour space at which the colour in question is most readily identified).

On the left of the grid the vertical scale 2 – 9 represents lightness intensity in ascending order while across the top the labels

5R – 10RP denote the colour of the chips, where Red (5R), Yellow (5Y), Green (5G), Blue (5B) and Purple (5P) represent five principal hues and the accompanying terms correspond to fifteen intermediate hues. For instance, between Red (5R) and Yellow (5Y) there exist three intermediary hues: 10R, 5YR, 10YR (see Figure Y below).

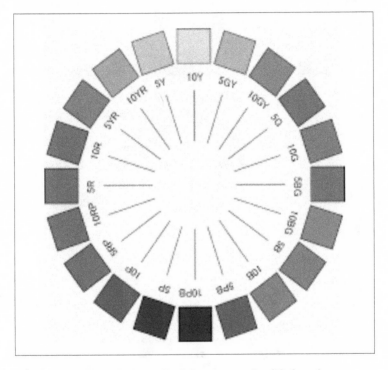

Figure Y – The Munsell colour system: a simplified version.

However, when these same chips are presented to Berinmo speakers – a language spoken in the Bitara and Kagiru villages located near the Sepik River in northeast Papua New Guinea – the following, somewhat divergent response pattern emerges:

	5R	10R	5YR	10YR	5Y	10Y	5GY	10GY	5G	10G	5BG	10BG	5B	10B	5PB	10PB	5P	10P	5RP	10RP
9	3	2	5	Wap	2			1	1				Wap	1	5	12	6	3		2
8				9	6	2	3										1			
7			2	5	4	4	1	1		2										
6		Mehi		2	1			2	3		1									
5	6	2		Wor	1			6	7	4	Nol	2	2						Mehi	3
4	19								5			3								11
3	2		Kel							1		1		1						1
2		1	1	3	4	6	12				2	1	4	3	4	4	Kel	2		

Figure Z – The colour space of native Berinmo speakers.

Though a number of noticeable differences clearly exist between the English and Berinmo colour space note how the most significant disparity between the two centres around the Berinmo amalgamation of the colours 'blue' and 'green' into the single colour category 'nol'.

The three evolutionary stages of black-and-white thinking

: :

TIMELINE (YEARS)	STAGE	DESCRIPTION
500,000,000 +	*Fight vs. Flight*	Predation has been around for donkeys' years. In fact, it predates donkeys by about 500,000 millennia, spearheading the advance of complex, diverse and increasingly sophisticated life forms from the unprecedented proliferation of species in the Cambrian era (c. 540–485 million years ago) to the emergence, first of mammals and eventually of primates, a few hundred million years later. But where there is predation there must also be defence – and during the course of evolutionary history these two opponent processes have vied for ascendancy in a titanic Darwinian arms race in which the escalation, on the one hand, of the hunter's predatory smarts has led, on the other, to increased biological investment in defence strategy R&D. Given its ubiquity among complex vertebrates, and the fact that early and reliable recognition of a predator is a prerequisite for evasive action, it would seem highly probable that one

500,000,000 – 6,000,000	Us vs. Them	of the very first blueprints to emerge from our Defence Department's R&D laboratories was that for a quick and dirty, no-nonsense 'fear module' in the brain: a neural structure that, after a minimum of computations, was able – involuntarily, automatically and pre-consciously – to flag up and categorize stimuli related to recurrent and wide-ranging survival threats and to assign them immediate priority.
		Social animals live in groups because the opportunities for survival and reproduction are better than living alone. Chimpanzees, our closest living relatives, for example, live in fission–fusion* groups that average c.50 individuals. It's likely that our prehistoric ancestors who walked the earth around 6 million years ago cohabited in groups of similar size, and that, to ensure cohesion, they evolved a propensity for in-group favouritism. The precise mechanism underlying the development of this 'new' socio-cognitive software is unknown. But the likelihood is that it piggybacked on an ancient, much earlier adaptation – around 500 million years old and observed across many phylogenetic taxa including fish, birds, reptiles and mammals – for social exclusion/territoriality. Moreover, evidence suggests that ever since the days of our foraging forebears our brains have acquired specialized neurocognitive technology explicitly designed to both detect and

---

* In ethology, a fission–fusion society is one in which the size and composition of the social group is fluid and subject to regular and consistent change over a set period of time. For instance, a troop of chimpanzees might sleep in one place during the hours of darkness (*fusion*) but might split up during the day to forage in small groups (*fission*).

		track shifting coalitions and alliances within group settings – a necessary requirement for hunter-gatherer communities even today (including, of course, online communities of *social* hunter-gatherers).
100,000	*Right vs. Wrong*	As stated above, our earliest hominid ancestors lived in groups that consisted, on average, of around 50 individuals. Extrapolating from the size of existing hunter-gatherer communities, it's a reasonable assumption that more recent Paleolithic hominids – those who lived around 100,000 years ago – cohabited in bands of a few hundred individuals. As this social 'bandwidth' increased over the course of human evolution, the pressure to maintain group solidarity would have intensified and natural selection would, in turn, have selected for those adaptations most effective in reinforcing it. Morality, then, may well have evolved in these early communities of 100 to 200 individuals as an instrument of social control: as a means of deterring widespread aberrant self-interest, of managing and resolving conflict, and of maximizing group cohesion. It is striking, for instance, just how much accord there exists between cultures, both past and present, as to the nature and corollaries of virtue: deception, aggression, vindictiveness and selfishness are all universally frowned upon, while courage, modesty, leadership and cooperation are encouraged across the board. Such norms and values pre-date the existence of a just and moral God by quite some considerable

time – and, in all likelihood,
language, too – suggesting that
their roots may instead be discerned
in the acquisition of behavioural
patterns specifically designed to
safeguard the integrity of the group.
In short, good habits died hard
because if *they* died we died with
them.

Black-and-white thinking
through the centuries

: :

It is possible, within the traditions of Western thought, to identify cultural and historical differences in black-and-white thinking over the centuries which would appear to follow a loosely recurring pattern of binary boom and bust. In Medieval Europe, for example, the structure of society in terms of employer relations and endemic social hierarchies – the combined economic and interpersonal framework of which formed the transactional basis of feudalism – was considered immutable. Each link in the 'great chain of being' was regarded as being of equal and utmost importance – St Augustine's notion that God had assigned each individual a fixed place in the social order was deemed infallible – and anyone questioning his or her station within the community was reminded of the saint's studiously unempowering metaphor of a dissatisfied finger desiring to be an eye.

In contrast, the Age of Enlightenment (1715 to 1789) is viewed by historians as being very much an age of intellectual and philosophical rebellion, a period defined by the pursuit of individual liberty and the demise of absolute monarchies, along, of course, with the championing of religious tolerance in the face of ideological orthodoxy. Corresponding appeals to logic, reason and empirical observation as means of scholastic inquiry and of gaining understanding and knowledge – the scientific method was born during this time – paved the way for a different kind of order, a naturalistic, reductionist order that carried right the way through into the Victorian era and is both epitomized and preserved by the Victorian predilection for collection and classification. (There was even a school of thought in some echelons of Victorian society, predominantly the upper and middle classes, that certain rooms within the home could ideally be used for just the one specific function. One couldn't, for example, read in the bedroom, one could only sleep. One couldn't play games in the kitchen, one could only cook.)

This order was itself usurped, towards the end of the nineteenth century, by the modernist creed of flux, innovation and experimentation, fuelled by a growing disaffection with Victorian moral principles and sociocultural conventions. Inevitably, given its ancient philosophical roots in the intellectual mores of the eighteenth century – though the mind might achieve enlightenment through reason it might also, according to modernists, achieve it through *un*reason – modernism was not without its own set of ideological dress codes and conceptual house rules. It insisted, for example, on a clear divide between art and popular culture and modernists, in their own way – albeit employing different methodologies within different moral, psychological and aesthetic parameters – were just as intent on divining unified meanings and universal truths as were the rationalists and empiricists of the Enlightenment and the Age of Reason.

It was with the shift from modernism to postmodernism in the 1970s, and the attendant rise of social constructionism and 'incredulity towards metanarratives', as the French philosopher Jean-François Lyotard once famously described it, that the lines of intellectual inquiry were blurred once again, and the nebulous genies of subjectivism, relativism and pluralism were released from their smoke-filled bottles.

As the comedic early twentieth-century military figure General Stumm von Bordwehr – who is anything but *stumm* (German for 'mute') – puts it in Robert Musil's novel *The Man Without Qualities*: 'Somehow or another, order, once it reaches a certain stage, calls for bloodshed.'

The essentials of essentialism

Polarization, overstatement and hyperbole have always featured prominently in the cut and thrust of political discourse, from Ancient Greece to New Labour, from Gandhi to Genghis Khan. But it's possible that other psychological forces relating to black-and-white thinking were also at play in the antipathy directed at Tony Blair's ostensibly sensible testing strategy from certain sectors of the British public.

Have you ever wondered how celebrity auctions work? Freddie Mercury's 'Invisible Man' sunglasses. Kurt Cobain's 'MTV unplugged' cardigan. An Elvis Presley pill bottle. All of these items have gone under the hammer in recent years, fetching upward of five-figure sums. Why? Cobain's cardigan was old and manky. Elvis's pill bottle was empty.

What's the appeal?

A clue lies on the flip side. A student of mine once conducted a study in which she asked participants how comfortable they'd be receiving an organ donation from the Yorkshire Ripper. Not very was the answer. By the same token, we don't buy Gary Glitter records any more. People contaminate or infuse – depending on how we see them – inanimate objects in the same way that germs might contaminate a door handle or a scent infuses a room. A part of 'them' lives on in 'what was theirs' and such random artefacts encapsulate and disseminate their essence.

Such persistent belief in an imperishable psychological soul and its ability to take up residence in eyewear, knitwear and the assorted paraphernalia of drug-dispensing descends from the philosophical notion of essentialism. Essentialism contends that there are intangible qualities in everything that's out there, that give stuff, as Aristotle conceived of it over two thousand years ago now, its 'substance', and that, as the American linguist and philosopher George Lakoff put it more recently, 'make the thing what it is, and without which it would be not that kind of thing'.

But such essences are eclectic and pervasive. They lurk not just in what we wear but in what we say. It's a well-known precept of influence theory that the power of a persuasive message resides in three separate domains: the content of the message itself, the target audience (i.e. the person or persons for whom the message is intended), and the source of the message (i.e. who is delivering it).

Obviously, if the message is inappropriate or the audience is non-receptive, then influence is less likely to occur. But if the message is good and the audience is sympathetic, influence effects are still likely to be compromised if the source of the message is perceived to lack credibility. Who the message is from – who 'wears it', as it were – contaminates (or infuses) what it *is*.

Indeed, if essentialism pervades the art of persuasive messaging, then it can also seep into the fabric of language itself. During the latter part of February 2020 as concerns over Coronavirus began to escalate, a survey of over seven hundred beer drinkers in the US conducted by 5W Public Relations revealed that 38 per cent of Americans wouldn't buy Corona beer 'under any circumstances' and 14 per cent said that they wouldn't order a Corona in public. Another survey, conducted by YouGov, found that consumers' intent to purchase the brand had fallen to its lowest level in two years. Consumption of Furlough Merlot and Quarantinis, on the other hand – experimental 'locktails' mixed with random, leftover ingredients from boozy, long-forgotten holidays on the Costas and Greek islands – went through the roof.

References

Introduction

Andrew Sparrow, 'Coronavirus: UK over-70s to be asked to stay home "within weeks", Hancock says', *Guardian*, 15 March 2020. (https://www.theguardian.com/world/2020/mar/15/coronavirus-uk-over-70s-to-be-asked-to-self-isolate-within-weeks-hancock-says)

BBC Sport, 'Caster Semenya: Fans urge athlete not to quit after her cryptic tweet'. (https://www.bbc.co.uk/sport/athletics/48131074)

Schaeffer, A. A., 'Reactions of Ameba to Light and the Effect of Light on Feeding', *Biological Bulletin* 32(2), 45–74 (1917).

Christian Jarrett, 'Psychology: How many senses do we have?', *BBC Future*, 19 November 2014. (https://www.bbc.com/future/article/20141118-how-many-senses-do-you-have)

James Vincent, 'Facebook introduces more than 70 new gender options to the UK: "We want to reflect society"', *Independent*, 27 June 2014. (https://www.independent.co.uk/life-style/gadgets-and-tech/facebook-introduces-more-than-70-new-gender-options-to-the-uk-we-want-to-reflect-society-9567261.html)

For up-to-the-minute analysis on Spotify genres, see: http://everynoise.com.

Alexis C. Madrigal, 'How Netflix reverse-engineered Hollywood', *The Atlantic*, 2 January 2014. (https://www.theatlantic.com/technology/archive/2014/01/how-netflix-reverse-engineered-hollywood/282679/)

Gordon Rayner & Jack Taperell, 'London 2012 Olympics: How Danny Boyle got 60,000 fans to keep mum over opening ceremony dress

rehearsal', *Telegraph*, 24 July 2012. (https://www.telegraph.co.uk/sport/
olympics/london-2012/9423842/London-2012-Olympics-How-Danny-
Boyle-got-60000-fans-to-keep-mum-over-opening-ceremony-dress-
rehearsal.html)

Whitehead, Alfred North, *Science and the Modern World* (Cambridge: Cambridge University Press, 1926).

Chapter 1: The Categorization Instinct

Ledford, H., 'The tell-tale grasshopper: can forensic science rely on the evidence of bugs?' *Nature* (2007). (doi:10.1038/news070618–5) (https://www.nature.com/news/2007/070619/full/news070618-5.html)

Quinn, P. C., Eimas, P. D. & Rosenkranz, S. L., 'Evidence for representations of perceptually similar categories by 3-month-old and 4-month-old infants', *Perception* 22, 463–75 (1993).

Oakes, L. M. & Ribar R. J., 'A comparison of infants' categorization in paired and successive presentation familiarization tasks', *Infancy* 7, 85–98 (2005).

Spalding L. & Murphy L., 'The effects of background knowledge on category construction', *Journal of Experimental Psychology: Learning, Memory and Cognition* 22, 525–38 (1996).

For how categorization can vary according to different goals, see: Medin, D. L., Lynch, E. B., Coley, J. D. & Atran, S., 'Categorization and reasoning among tree experts: Do all roads lead to Rome?' *Cognitive Psychology* 32, 4996 (1997).

For the cow/pig study, see: Gelman, S. & Wellman, H., 'Insides and essences', *Cognition* 38, 214–44 (1991).

For the porcupine/cactus study, see: Keil, F. C., *Concepts, kinds, and cognitive development* (Cambridge, MA: MIT Press, 1989).

For more on the concept of emotional granularity, see: Drake Baer, 'People with high "emotional granularity" are better at being sad', *The Cut*, 13 June 2016. (https://www.thecut.com/2016/06/people-with-high-emotional-granularity-are-better-at-being-feeling-things.html)

For an encyclopedic account of all things categorical, see: Harnad, Steven (Ed.), *Categorical perception: The groundwork of cognition* (Cambridge: Cambridge University Press, 1987).

Chapter 2: A Heap of Trouble

For a history of Parkrun from its inception right up to the present, see: https://blog.parkrun.com/uk/2019/10/02/a-look-back-at-15-years-of-parkrun/

Stephen Morris, 'Council condemned for charging runners to use park', *Guardian*, 13 April 2016. (https://www.theguardian.com/lifeandstyle/2016/apr/13/stoke-gifford-parkrun-council-condemned-charge-paula-radcliffe)

'Talking point: Should we be charged for running in a park?' *Sports Management* 119, 2 May 2016. (https://www.sportsmanagement.co.uk/Sports-features/sports-management-magazine/Should-we-be-charged-for-running-in-a-park/30778)

Tim Shipman & Caroline Wheeler, 'Coronavirus: Ten days that shook Britain – and changed the nation for ever', *Sunday Times*, 22 March 2020. (https://www.thetimes.co.uk/article/coronavirus-ten-days-that-shook-britain-and-changed-the-nation-for-ever-spz6sc9vb?shareToken=66410e6a661ccba25dfc5886be410004)

Larisa Brown, Downing Street fury over 'let OAPs die' accusation: 'No 10 angrily denies claims that Dominic Cummings said pensioners should be sacrificed to coronavirus to protect the economy', *Mail Online*, 22 March 2020. (https://www.dailymail.co.uk/news/article-8140795/No10-denies-claims-Dominic-Cummings-said-pensioners-sacrificed-virus-protect-economy.html)

Seuren, P. A. M., 'Eubulides as a 20th-century semanticist', *Language Sciences* 27(1), 75–95 (2005).

For more on the Sorites paradox and related themes, see: Van Deemter, Kees, *Not exactly: in praise of vagueness* (Oxford: Oxford University Press, 2010).

For an early take on the phenomenon of the just noticeable difference, see: Newman, E. B., 'The validity of the Just Noticeable Difference as a unit of psychological magnitude', *Transactions of the Kansas Academy of Science* 36, 172–5 (1933).

Plato, *The Republic*, translated by H. D. P. Lee (London: Penguin Classics, 1955).

Quinn, W., 'The paradox of the self-torturer', *Philosophical Studies* 59, 79–90 (1990).

Chapter 3: When Categories Collide

Ackrill, John, *Aristotle, Categories and De Interpretatione* (Oxford: Clarendon Press, 1963).

Aristotle, 'Categories', in Barnes, J. (Ed.), *The complete works of Aristotle*, translated by Ackrill, J. L. (Princeton: Princeton University Press, 2014).

Maloney, Kellie, *Frankly Kellie: Becoming a woman in a man's world* (London: Blink, 2015).

Wittgenstein, Ludwig, *Philosophical Investigations*, 2nd ed., translated by G. E. M. Anscombe (Oxford: Basil Blackwell Ltd, 1958).

Keiligh Baker, 'Schools are told to call transgender children "zie" rather than "he" or "she" in case they cause offence', *Mail Online*, 17 July 2016. (https://www.dailymail.co.uk/news/article-3694194/Schools-told-call-transgender-children-zie.html)

Rosa Doherty, 'Transgender woman who grew up in Charedi community abandons fight to see her children', *The Jewish Chronicle*, 21 January 2020. (https://www.thejc.com/news/uk-news/transgender-woman-ends-fight-to-see-her-children-1.495683)

DFA Law, 'Transgender woman objects to being called a father', *DFA Law Family Law, General News*, 19 May 2015. (https://www.dfalaw.co.uk/transgender-woman-objects-to-being-called-a-father/)

Jamie Bullen, 'BMA warns against calling pregnant women "expectant mothers" as it "may offend intersex men"', *Evening Standard*, 29 January 2017. (https://www.standard.co.uk/news/health/bma-warns-against-using-expectant-mothers-for-pregnant-women-as-it-may-offend-intersex-men-a3452736.html)

Ellie Cambridge, 'Don't do it! First man to give birth in UK warns others not to try it because it's "really hard"', *Sun*, 9 January 2019. (https://www.thesun.co.uk/news/8152999/first-man-give-birth-uk-warns-really-hard/)

Ione Wells, 'From "hin" to "zie": how pronouns are moving beyond gender', *Evening Standard*, 20 July 2016. (https://www.standard.co.uk/lifestyle/london-life/from-hin-to-zie-how-pronouns-are-moving-beyond-gender-a3300186.html)

Rosch Heider, E., 'Probabilities, sampling and ethnographic method: The case of Dani colour names', *Man* 7, 448–66 (1972).

Rosch Heider, E. & D. C. Olivier, 'The structure of the color space for naming and memory in two languages', *Cognitive Psychology* 3, 337–54 (1972).

Rosch, E. H., 'Natural categories', *Cognitive Psychology* 4, 328–50 (1973).

Rosch, E., 'Cognitive Representations of Semantic Categories', *Journal of Experimental Psychology: General* 104(3), 192–233 (1975).

Berlin, Brent & Paul Kay, *Basic Color Terms: Their universality and evolution* (Berkeley and Los Angeles: University of California Press, 1969).

David Conn, 'Hillsborough: David Duckenfield found not guilty of manslaughter', *Guardian*, 28 November 2019. (https://www.theguardian.com/uk-news/2019/nov/28/hillsborough-david-duckenfield-verdict)

Faúndes, Aníbal & José S. Barzelatto, *The human drama of abortion: A global search for consensus* (Nashville, TN: Vanderbilt University Press, 2006).

Kitty Holland, 'How the death of Savita Halappanavar revolutionised Ireland', *Irish Times*, 28 May 2018. (https://www.irishtimes.com/news/social-affairs/how-the-death-of-savita-halappanavar-revolutionised-ireland-1.3510387)

Gethin Chamberlain, ' "Change your abortion law to save lives" grieving father tells Irish PM', *Observer*, 17 November 2012. (https://www.theguardian.com/world/2012/nov/17/ireland-abortion-case-father)

Borges, Jorge Luis, 'The analytical language of John Wilkins', in *Other Inquisitions, 1937–1952*, translated by Ruth L. C. Simms (Austin, Texas: University of Texas Press, 1964).

Chapter 4: The Dark Side of Black and White

Luke Traynor, 'Welsh "mountain" downgraded to hill after experts say it's not big enough', *Daily Mirror*, 7 August 2015. (https://www.mirror.co.uk/news/uk-news/welsh-mountain-downgraded-hill-after-6211672)

Nuttall, John & Anne, *The Mountains of England & Wales*, Vol 1, 3rd ed.: *Wales* (Milnthorpe, Cumbria: Cicerone, 2009).

Nuttall, John & Anne, *The Mountains of England & Wales*, Vol 2, 3rd ed.: *England* (Milnthorpe, Cumbria: Cicerone, 2008).

Rubin, Richard E., *Foundations of library and information science*, 4th ed. (London: Facet, 2015).

Iyengar, S. S. & Lepper, M. R., 'When choice is demotivating: Can one desire too much of a good thing?' *Journal of Personality and Social Psychology* 79(6), 995–1006 (2000). (https://doi.org/10.1037/0022-3514.79.6.995)

Schwartz, Barry, *The paradox of choice: Why less is more* (New York: Harper Perennial, 2004).

Feldman Barrett, Lisa, *How emotions are made: The secret life of the brain* (New York: Houghton Mifflin Harcourt, 2017).

Miller, G.A., 'The magical number seven, plus or minus two: Some limits on our capacity for processing information', *Psychological Review* 63(2), 81–97 (1956).

Cowan, N., 'George Miller's magical number of immediate memory in retrospect: Observations on the faltering progression of science', *Psychological Review* 122(3), 536–41 (2015). (doi:10.1037/a0039035)

Preston, C. C. & Colman, A. M., 'Optimal number of response categories in rating scales: reliability, validity, discriminating power, and respondent preferences', *Acta Psychologica* 104, 1–15 (2000).

McCrae, R. R. & John, O. P., 'An introduction to the Five-Factor Model and its applications', *Journal of Personality* 60(2), 175–215 (1992). (doi:10.1111/j.1467-6494.1992.tb00970.x)

Hofstede, Geert, 'Dimensionalizing cultures: The Hofstede model in context', *Online Readings in Psychology and Culture*, 2(1) (2011). (http://dx.doi.org/10.9707/2307-0919.1014)

Beebe-Center, J. G., Rogers, M. S. & O'Connell, D. N., 'Transmission of information about sucrose and saline solutions through the sense of taste', *Journal of Psychology* 39, 157–160 (1955).

Luchian, S. A., McNally, R. J. & Hooley, J. M., 'Cognitive aspects of non-clinical obsessive-compulsive hoarding', *Behaviour Research and Therapy* 42, 1657–62 (2007).

Wincze, J. P., Steketee, G. & Frost, R. O., 'Categorization in compulsive hoarding', *Behaviour Research and Therapy* 45(1), 63–72 (2007).

Grisham, J. R., Norberg, M. M., Williams, A. D., Certoma, S. P. & Kadib, R., 'Categorization and cognitive deficits in compulsive hoarding', *Behaviour Research and Therapy* 48(9), 866–72 (2010).

For the definitive read on hoarding, see: Frost, Randy O., & Steketee, G. (Eds), *The Oxford handbook of acquiring and hoarding* (Oxford: Oxford University Press, 2014).

Reed, G. F., 'Underinclusion – A characteristic of obsessional personality disorder: I', *British Journal of Psychiatry* 115, 781–5 (1969). (doi: 10.1192/bjp.115.524.781)

Reed, G. F., 'Underinclusion – A characteristic of obsessional personality disorder: II', *British Journal of Psychiatry* 115, 787–90 (1969). (doi: 10.1192/bjp.115.524.787)

Nelson, Todd D. (Ed.), *Handbook of prejudice, stereotyping and discrimination*, 2nd ed. (New York: Psychology Press, 2016).

Chapter 5: The Viewfinder Principle

For another account of the Finnish farmer story, see: DeMello, Anthony, *Awareness* (London: HarperCollins, 1990).

Rowena Mason & Kate Proctor, 'Coronavirus: UK offers self-employed 80% of earnings', *Guardian*, 26 March 2020. (https://www.theguardian.com/world/2020/mar/26/coronavirus-uk-offers-self-employed-80-of-earnings)

Butcher, Pat, *The perfect distance – Ovett and Coe: the record breaking rivalry* (London: Weidenfeld & Nicolson, 2004).

Davidson, Ian, *The French revolution: From enlightenment to tyranny* (London: Profile Books, 2017).

Lucy Pasha-Robinson, 'Labour MP says she could never be friends with a Tory: "Whatever type they are, I have absolutely no intention of being friends with any of them",' *Independent*, 24 August 2017. (https://www.independent.co.uk/news/uk/politics/laura-pidcock-labour-mp-tory-never-friends-conservative-party-a7910156.html)

For more on 'Marilyn Einstein', see: https://www.techtimes.com/articles/44206/20150404/marilyn-monroe-or-albert-einstein-optical-illusion-can-tell-if-you-need-glasses-or-not.htm

Chapter 6: The Complexity of Simplicity

Bachet, Claude-Gaspard, *Problèmes, plaisants et délectables qui se font par les nombres*, 1624 (London: Forgotten Books, 2019).

Tschoegl, A. E., 'The optimal denomination of currency', *Journal of Money, Credit and Banking* 29(4), 546–54 (1997).

Mohlin, E., 'Optimal categorization', *Journal of Economic Theory* 152, 356–81 (2014). (https://doi.org/10.1016/j.jet.2014.03.007)

Holmes, Jamie, *Nonsense: The power of not knowing* (New York: Crown, 2015).

Victoroff, Jeff & Arie W. Kruglanski (Eds), *Psychology of terrorism* (New York: Psychology Press, 2009).

Kruglanski, Arie W., *The psychology of closed mindedness* (New York: Psychology Press, 2004).

Webster, D. M. & Kruglanski, A. W., 'Cognitive and social consequences of the need for cognitive closure', *European Review of Social Psychology* 18, 133–73 (1997).

Kruglanski, A. W. & Webster, D. M., 'Motivated closing of the mind: "Seizing" and "freezing"', *Psychological Review* 103, 263–83 (1996).

Webster, D. M. & Kruglanski, A. W., 'Individual differences in need for cognitive closure', *Journal of Personality and Social Psychology* 67, 1049–62 (1994).

Darwin, C., *The origin of species by means of natural selection*, 6th ed. (London: John Murray, 1872).

Jaensch, Erich R., *Der Gegentypus: Psychologischanthropologische Grundlagen deutscher Kulturphilosophie ausgehend von dem was wir überwinden wollen* (Leipzig: Ambrosius Barth, 1938).

Frenkel-Brunswik, E., 'Intolerance of ambiguity as an emotional and perceptual personality variable', *Journal of Personality* 18, 108–43 (1949). (https://doi.org/10.1111/j.1467-6494.1949.tb01236.x)

Bieri, J., 'Cognitive complexity-simplicity and predictive behavior', *Journal of Abnormal and Social Psychology* 51, 263–8 (1955).

Crockett, W. H., 'Cognitive complexity and impression formation', in B. A. Maher (Ed.), *Progress in experimental personality research*, Vol. 2, 47–90 (New York: Academic Press, 1965).

Crockett, W. H., 'The organization of construct systems: The organization corollary', in Mancuso, J. C. & Adams-Webber, J. R. (Eds), *The construing person*, 62–95 (New York: Praeger, 1982).

Rafaeli-Mor, E., Gotlib, I. H. & Revelle, W., 'The meaning and measurement of self-complexity', *Personality and Individual Differences* 27, 341–56 (1999).

Vannoy, J. S., 'Generality of cognitive complexity-simplicity as a personality construct', *Journal of Personality and Social Psychology* 2, 385–96 (1965).

For more on the art and science of framing, see: Lakoff, George, *Women, fire and dangerous things: What categories reveal about the mind* (Chicago: University of Chicago Press, 1987); and Lakoff, George, *Don't think of an elephant: Know your values and frame the debate* (White River Junction, VA: Chelsea Green, 1987).

Tom Embury-Dennis, 'Trump says military is "waiting for" migrant caravan after warning of "invasion" across US border', *Independent*, 29 October 2018.(https://www.independent.co.uk/news/world/americas/trump-migrant-caravan-tweet-mexico-border-tracker-location-honduras-tapanatepec-a8607031.html)

Chapter 7: The Rainbow That Might Have Been

Kelley, H. H., 'The warm–cold variable in first impressions of persons', *Journal of Personality* 18, 431–9 (1950).

Bargh, J. A. & Williams, L. E., 'Experiencing physical warmth promotes interpersonal warmth', *Science* 322 (5901), 606–7 (2008). (doi: 10.1126/science.1162548); however, see also: Chabris, C. F., Heck, P. R., Mandart, J., Benjamin, D. J. & Simons, D. J., 'No evidence that experiencing physical warmth promotes interpersonal warmth: Two failures to replicate Williams and Bargh (2008)', *Social Psychology* 50(2), 127–32 (2019). (https://doi.org/10.1027/1864-9335/a000361)

Loftus, E. F. & Palmer, J. C., 'Reconstruction of automobile destruction: An example of the interaction between language and memory', *Journal of Verbal Learning and Verbal Behavior* 13(5), 585–9 (1974). (doi:10.1016/S0022-5371(74)80011-3)

Lifton, Robert J., *Thought reform and the psychology of totalism: A study of 'brainwashing' in China* (Chapel Hill, NC: University of North Carolina Press, 1961, 1989).

Orwell, George, 'Politics and the English language', *Horizon* 13(76), 252–65 (1946).

Whorf, B. L., 'Science and linguistics', *Technology Review* 42, 227–31, 247–8 (1940), reprinted in Carroll, John B., Levinson, S. C. & Lee, P. (Eds), *Language, thought, and reality: Selected writings of Benjamin Lee Whorf* (Cambridge, MA: The Technology Press of MIT/New York: Wiley, 1956, 2012).

Pinker, Steven, *The language instinct: how the mind creates language* (New York: William Morrow, 1994).

Koerner, E. F. K., 'The Sapir-Whorf hypothesis: A preliminary history and a bibliographical essay', *Journal of Linguistic Anthropology* 2(2), 173–98 (1992).

Wolff, P. & Holmes, K. J., 'Linguistic relativity', *Wiley Interdisciplinary Reviews: Cognitive Science* 2(3), 253–65 (2011).

Jraissati, Y., 'On color categorization: Why do we name seven colors in the rainbow?', *Philosophy Compass* 9(6), 382–91 (2014). (https://doi.org/10.1111/phc3.12131)

Davidoff, J., Davies, I. R. L. & Roberson, D., 'Is color categorisation universal? New evidence from a stone-age culture', *Nature* 398, 203–04 (1999).

Roberson, D., Davidoff, J., Davies, I. R. L. & Shapiro, L. R., 'Color categories: Evidence for the cultural relativity hypothesis', *Cognitive Psychology* 50, 378–411 (2004).

Roberson, D., Davies, I. R. L., & Davidoff, J., 'Color categories are not universal: Replications and new evidence from a Stone-age culture', *Journal of Experimental Psychology: General* 129, 369–98 (2000).

For more on the history of orange and other colours, see: Kastan, D. S. & Farthing, S., *On Color* (New Haven, CT: Yale University Press, 2018).

Wittgenstein, L., *Philosophical investigations* (New York: Macmillan, 1953).

Levvis, G. W., 'Why we would not understand a talking lion', *Between The Species* 8, 156–62 (1992).

Levinson, S. C., 'Language and Cognition: The Cognitive Consequences of Spatial Description in Guugu Yimithirr', *Journal of Linguistic Anthropology* 7(1), 98–131 (1997).

For more on how culture and context can influence the content and structure of thought, see: Ornstein, Robert & Paul Ehrlich, *New world–new mind: Moving toward conscious evolution* (New York: Doubleday, 1989).

Ahlberg, Per Erik, *Major events in early vertebrate evolution: palaeontology, phylogeny, genetics and development* (Washington, DC: Taylor & Francis, 2001).

For more on the conditions of early life on earth, see: Dawkins, Richard, *The ancestor's tale: A pilgrimage to the dawn of life* (London: Weidenfeld & Nicolson, 2004).

Geiger, Lazarus, *Ursprung und Entwicklung der menschlichen Sprache und Vernunft* (Stuttgart: JG Cotta, 1868).

Gladstone, William E., *Studies on Homer and the Homeric age* (Oxford: Oxford University Press, 1858).

For more on the relationship between language and thought (including colour perception), see: Deutscher, Guy, *Through the language glass: Why the world looks different in other languages* (London: William Heinemann, 2010).

Loreto, V., Mukherjee, A. & Tria, F., 'On the origin of the hierarchy of color names', *Proceedings of the National Academy of Science of the United States of America* 109(18), 6819–24 (2012). (doi: 10.1073/pnas.1113347109)

For our co-option of nature's warning colouration to accommodate modern-day needs, see: Forbes, Peter, *Dazzled and deceived: Mimicry and camouflage* (London: Yale University Press, 2011).

Ho, H.-N., Van Doorn, G. H., Kawabe, T., Watanabe, J., Spence, C., 'Colour-temperature correspondences: When reactions to thermal stimuli are influenced by colour', *PLoS One* 9(3) (2014). (doi: 10.1371/journal.pone.0091854)

Ho, H.-N., Iwai, D., Yoshikawa, Y., Watanabe, J., Nishida, S., 'Combining colour and temperature: A blue object is more likely to be judged as warm than a red object', *Scientific Reports* 4, Article 5527 (2014). (doi:10.1038/srep05527)

Rachel Holmes, 'Why blue is the costliest colour', *Guardian*, 17 April 2015. (https://www.theguardian.com/lifeandstyle/2015/apr/17/colour-blue-rich-divine-ancient-egyptians-virgin-mary)

Mark Molloy, 'Roger Federer settles the "green or yellow" tennis ball debate', *Telegraph*, 22 March 2018. (https://www.telegraph.co.uk/tennis/2018/03/22/roger-federer-settles-green-yellow-tennis-ball-debate/)

Steve Bird, 'Chocolate wars break out over the colour purple', *Telegraph*, 2 February 2019. (https://www.telegraph.co.uk/news/2019/02/02/chocolate-wars-break-colour-purple/)

Julia Day, 'Easy brand's future may not be orange,' *Guardian*, 16 August 2004. (https://www.theguardian.com/media/2004/aug/16/newmedia.marketingandpr)

Zoe Wood, 'Christian Louboutin wins ECJ ruling over red-soled shoes', *Guardian*, 12 June 2018. (https://www.theguardian.com/business/2018/jun/12/christian-louboutin-ecj-ruling-red-soled-shoes)

Chapter 8: The Frame Game

Susie Dent, 'X-rated make-up to make you blush', *Standard Issue*, 15 February 2016. (http://standardissuemagazine.com/lifestyle/x-rated-makeup-to-make-you-blush/?utm_content=bufferf2ea5&utm_medium=social&utm_source=twitter.com&utm_campaign=buffer)

Beukeboom, C., Tanis, M. & Vermeulen, I., 'The language of extraversion: Extraverted people talk more abstractly, introverts are more concrete', *Journal of Language and Social Psychology* 32(2), 191–201 (2012). (doi: 10.1177/0261927X12460844)

Christian Jarrett, 'The hidden ways your language betrays your character', *BBC Future*, 21 July 2017. (https://www.bbc.com/future/article/20170720-the-hidden-ways-your-language-betrays-your-character)

Lisa Feldman Barrett, 'Are you in despair? That's good', *New York Times*, 3 June 2016. (https://www.nytimes.com/2016/06/05/opinion/sunday/are-you-in-despair-thats-good.html?_r=0)

Kashdan, T. B., Feldman Barrett, L. & McKnight, P. E. 'Unpacking emotion differentiation: Transforming unpleasant experience by perceiving distinctions in negativity', *Current Directions in Psychological Science* 24(1), 10–16 (2015).

Wiking, M., *The little book of hygge: Danish secrets to happy living* (London: Penguin Life, 2016).

Frayer, D. W., Clarke, R. J., Fiore, I., Blumenschine, R. J., Pérez-Pérez, A., Martinez, L. M., Estebaranz, F., Holloway, R. & Bondioli, L., 'OH-65: The earliest evidence for right-handedness in the fossil record', *Journal of Human Evolution* 100, 65–72 (2016).

McManus, I. C., 'The history and geography of human handedness', in Sommer, I. & Khan, R. S. (Eds), *Language Lateralisation and Psychosis* (Cambridge: Cambridge University Press, 2009).

Hannah Fry, 'The mystery of why left-handers are so much rarer', *BBC Future*, 3 October 2016. (https://www.bbc.com/future/article/20160930-

the-mystery-of-why-left-handers-are-so-much-rarer?ocid=fbfut&fbclid=
IwAR12-t1vKn88ije7FvhClZXfLb57v439VnQXBKKS8QH-oTY1MF
byQJA7-u0&referer=https%3A%2F%2Fwww.facebook.com%2F)

For more on the science of handedness, see: McManus, Chris, *Right hand, left hand* (London: Weidenfeld & Nicolson, 2002).

For more on the field of embodied cognition, see: Clark, Andy, *Supersizing the mind: Embodiment, action, and cognitive extension* (Cambridge: Cambridge University Press, 2008); and Varela, Francisco J., Thompson, Evan & Rosch, Eleanor, *The embodied mind: Cognitive science and human experience* (Cambridge, MA: MIT Press, 1991).

Frank J. Ninivaggi, ' "Two-ness:" The mind's binary code', *Psychology Today*, 21 June 2015. (https://www.psychologytoday.com/gb/blog/envy/201506/two-ness-the-mind-s-binary-code)

For more on the (psychoanalytic) concept of two-ness, see: Bion, Wilfred R., *Second thoughts: Selected papers on psychoanalysis* (London: Karnac Books, 1967, 1993).

Ifrah, Georges, *The universal history of numbers: From prehistory to the invention of the computer*, translated by David Bellos, E. F. Harding, Sophie Wood & Ian Monk (New York: Wiley, 2000).

Gvozdanović, Jadranka (Ed.), 'Numeral types and changes worldwide', *Trends in linguistics: Studies and monographs* 118 (Berlin: Mouton de Gruyter, 1999).

Mercier, H. & Sperber, D., 'Why do humans reason? Arguments for an argumentative theory', *Behavioral and Brain Sciences* 34(2), 57–74 (2011). (http://dx.doi.org/10.1017/S0140525X10000968)

Dor, Daniel, *The instruction of imagination: Language as a social communication technology* (New York: Oxford University Press, 2015).

Dor, D., 'The role of the lie in the evolution of human language', *Language Sciences* 63, 44–59 (2017). (http://dx.doi.org/10.1016/j.langsci.2017.01.001)

Aristotle & George A. Kennedy, *On rhetoric: a theory of civic discourse* (New York: Oxford University Press, 1991).

Kahneman, Daniel, *Thinking, fast and slow* (New York: Farrar, Straus & Giroux, 2011).

Kahneman, D. & Tversky, A., 'Choices, values, and frames', *American Psychologist* 39(4), 341–50 (1984). (doi:10.1037/0003-066X.39.4.341)

Tversky, A. & Kahneman, D., 'The framing of decisions and the psychology of choice', *Science* 211(4481), 453–8 (1981). (doi:10.1126/science.7455683)

Kahneman, D., Knetsch, J. L., & Thaler, R. H., 'Anomalies: The endowment effect, loss aversion, and status quo bias', *Journal of Economic Perspectives* 5(1), 193–206 (1991).

Kahneman, D., Knetsch, J. L., & Thaler, R. H., 'Experimental tests of the endowment effect and the Coase theorem', *Journal of Political Economy* 98(6), 1325–48 (1990).

See the following link for a similar take on Claire Smith's inspirational story. Claire heard the story second-hand so the tale of Sakena Yacoobi may well have been the original version. https://www.mcgrawprize.com/press/2016/09/bringing-access-and-education-to-afghanistan-sakena-yacoobi-story/

For the condom study, see: Linville, P. W., Fischer, G. W. & Fischoff, B., 'AIDS risk perceptions and decision biases', in Pryor, J. B. & Reeder, G. D. (Eds), *The social psychology of HIV infection*, 5–38 (Hillsdale, NJ: Lawrence Erlbaum Associates, 1993).

For the meat study, see: Sanford, A. J., Fay, N., Stewart, A. & Moxey, L., 'Perspective in statements of quantity, with implications for consumer psychology', *Psychological Science* 13, 130–4 (2002).

For the salary/inflation study, see: Kahneman, D., Knetsch, J. L. & Thaler, R. H., 'Fairness as a constraint on profit seeking: Entitlements in the market', *American Economic Review* 76(4), 728–41 (1986).

For the penalty shootout study, see: Jordet, G. & Hartman, E., 'Avoidance motivation and choking under pressure in soccer penalty shootouts', *Journal of Sport & Exercise Psychology* 30, 450–7 (2008).

Dominic Cummings, 'How the referendum was won', *Spectator*, 9 January 2017.(https://www.spectator.co.uk/article/dominic-cummings-how-the-brexit-referendum-was-won)

Amsterdam-Advisor.com, 'Coffee shops in Amsterdam: Why are the Amsterdam cannabis cafes allowed?' (http://www.amsterdam-advisor.com/coffee-shops-in-amsterdam.html)

Oliver Holmes, 'Rodrigo Duterte vows to kill 3 million drug addicts and likens himself to Hitler', *Guardian*, 1 October 2016. (https://www.theguardian.

com/world/2016/sep/30/rodrigo-duterte-vows-to-kill-3-million-drug-addicts-and-likens-himself-to-hitler)

Shafir, E., 'Choosing versus rejecting: Why some options are both better and worse than others', *Memory & Cognition* 21(4), 546–56 (1993).

Richards, Keith, *Life* (New York: Little, Brown and Company, 2010).

For more on latitudes of acceptance and rejection, see: Hovland, Carl I. & Sherif, Muzafer, *Social judgment: Assimilation and contrast effects in communication and attitude change* (New Haven, CT: Yale University Press, 1961); Sherif, Carolyn W., Sherif, Muzafer & Nebergall, Roger Ellis, *Attitude and attitude change: The social judgement-involvement approach* (Philadelphia: W.B. Saunders Company, 1965); and Griffin, Emory A., *A first look at communication theory* (New York: McGraw Hill, 2011).

For the Shepard illusion: Shepard, Roger, *Mind sights: Original visual illusions, ambiguities, and other anomalies* (New York: WH Freeman and Company, 1990).

For the vertical–horizontal illusion: Winch, W. H., 'The vertical–horizontal illusion in school children', *British Journal of Psychology* 2, 220–5 (1907).

For more on visual illusions in general, see: Robinson, J. O., *The psychology of visual illusion* (New York: Dover, 1972).

Declan Cashin, 'Would these floating 3D zebra crossings slow you down?' *BBC* website, 1 November 2017. (https://www.bbc.co.uk/bbcthree/article/3e316454-02b6-4b1a-8369-da71a2377f25)

"A while back when I was at the University of Cambridge I ran a study in which I monitored the speech patterns of forty undergraduate students for an hour a day for a week . . ." and *"To illustrate, let me tell you about another simple study I conducted . . . Each group was then handed a list of ten adjectives that they had to slot into the course of their everyday conversation a total of five times a day for a week . . ."* These two studies are unpublished and were conducted merely with the aim of generating pilot data. Replications welcome! The former study is foreshadowed by a similar take on black-and-white thinking by Summer Beretsky. See: Summer Beretsky, 'Cognitive distortion: How does black-and-white thinking hurt us?', *PsychCentral*, 18 May 2009. (https://psychcentral.com/blog/cognitive-distortion-how-does-black-and-white-thinking-hurt-us#1)

Chapter 9: Where There's a Why There's a Way

Dutton, Kevin, *Flipnosis: The art of split-second persuasion* (London: William Heinemann, 2010).

Cialdini, Robert B., *Influence: The psychology of persuasion* (New York, NY: William Morrow, 1984).

David Chiu, 'Jonestown: 13 things you should know about cult massacre', *Rolling Stone*, 17 November 2017. (https://www.rollingstone.com/culture/culture-features/jonestown-13-things-you-should-know-about-cult-massacre-121974/)

William Rubel, 'History of the Stone Soup folktale from 1720 to now', *Stone Soup*, September 2015. (https://stonesoup.com/about-the-childrens-art-foundation-and-stone-soup-magazine/history-of-the-stone-soup-story-from-1720-to-now/)

Cheung, V. K. M., Harrison, P. M. C., Meyer, L., Pearce, M. T., Haynes, J.-D. & Koelsch, S., 'Uncertainty and surprise jointly predict musical pleasure and amygdala, hippocampus, and auditory cortex activity', *Current Biology* 29(23), 4084–92.e4 (2019). (doi: 10.1016/j.cub.2019.09.067)

For Dr Steven Taylor's Coronavirus interview, see: https://www.independent.co.uk/news/world/americas/coronavirus-toilet-paper-panic-buying-covid-19-uk-australia-a9403351.html

Taylor, Steven, *The psychology of pandemics: Preparing for the next global outbreak of infectious disease* (Newcastle: Cambridge Scholars Publishing, 2019).

Experts recommend 'latte levy' of 25p per cup, *BBC News* website, 17 July 2019. (https://www.bbc.co.uk/news/uk-scotland-49024433)

'Carrier bags: why there's a charge', Department for Environment, Food & Rural Affairs website. (https://www.gov.uk/government/publications/single-use-plastic-carrier-bags-why-were-introducing-the-charge/carrier-bags-why-theres-a-5p-charge)

Poortinga, W., *Results of a field experiment to reduce coffee cup waste.* Summary report to Bewley's Tea & Coffee UK Ltd (Cardiff: Welsh School of Architecture & School of Psychology, Cardiff University, 2017).

Karmarkar, U. R. & Tormala, Z. L., 'Believe me, I have no idea what I'm talking about: The effects of source certainty on consumer involvement and persuasion', *Journal of Consumer Research* 36(6), 1033–49 (2010).

De Berker, A. O., Rutledge, R. B., Mathys, C., Marshall, L., Cross, G. F., Dolan, R. J. & Bestmann, S., 'Computations of uncertainty mediate acute stress responses in humans', *Nature Communications* 7, 10996 (2016). (https://doi.org/10.1038/ncomms10996)

'Off-licences added to list of "essential" retailers', *BBC News* website, 25 March 2020. (https://www.bbc.co.uk/news/business-52033260)

Heikkilä, M., 'Finland taps social media influencers during coronavirus crisis', *Politico*, 31 March 2020. (https://www.politico.eu/article/finland-taps-influencers-as-critical-actors-amid-coronavirus-pandemic/)

Chris Ricotta, 'Coronavirus: US gun sales shatter records as shops selling firearms are declared essential', *Independent*, 2 April 2020. (https://www.independent.co.uk/news/world/americas/gun-sales-us-background-checks-coronavirus-firearms-trump-record-march-a9442861.html)

Chapter 10: Supersuasion

John T. Bennett, 'Coronavirus: Trump bans travel from Europe to the US but exempts UK', *Independent*, 12 March 2020. (https://www.independent.co.uk/news/world/americas/us-politics/coronavirus-us-travel-ban-trump-europe-travel-uk-speech-countries-latest-a9396086.html)

For an accessible but comprehensive account of Brexit, see: O'Rourke, Kevin, *A short history of Brexit: From Brentry to Backstop* (London: Pelican, 2019).

Naji, Abu Bakr, *The management of savagery: The most critical stage through which the umma will pass*, translated by William McCants (Cambridge, MA: John M. Olin Institute for Strategic Studies, Harvard University, 2004).

Wood, B. A., 'Evolution of australopithecines', in Stephen Jones, Robert D. Martin & David R. Pilbeam (Eds), *The Cambridge encyclopedia of human evolution* (Cambridge: Cambridge University Press, 1994).

For more on how morality evolved to consolidate societal relationships and to preserve group cohesion, see: Churchland, P., 'Why do we care?', *New Scientist*, 28 September 2019, 44–7.

For more on the relationship and interaction between morality and group identity, see: Greene, Joshua, *Moral tribes: Emotion, reason, and the gap between us and them* (London: Atlantic, 2013).

For more on the relationship between morality and emotion, see: Haidt, Jonathan, *The righteous mind: Why good people are divided by politics and religion* (London: Penguin, 2012).

'Syria war: The online activists pushing conspiracy theories', *BBC News* (trending) website, 19 April 2018. (https://www.bbc.co.uk/news/blogs-trending-43745629)

Holmes, J., 'The Republican party has drowned itself in Trumpism. Lyin-Comey.com proves it', *Esquire*, 12 April 2018. (https://www.esquire.com/news-politics/a19753363/lyincomey-website-trump-republican-party/)

For more on the Marjory Stoneman Douglas school shooting, see: https://www.washingtonpost.com/news/morning-mix/wp/2018/02/21/i-am-not-a-crisis-actor-florida-teens-fire-back-at-right-wing-conspiracy-theorists/

David Roberts, 'Donald Trump and the rise of tribal epistemology', *Vox*, 19 May 2017. (https://www.vox.com/policy-and-politics/2017/3/22/14762030/donald-trump-tribal-epistemology)

Hastorf, A. H. & Cantril, H., 'They saw a game: a case study', *Journal of Abnormal and Social Psychology*, 49(1), 129–34 (1954). (https://doi.org/10.1037/h0057880)

Kahan, D. M., Hoffman, D. A., Braman, D., Evans, D. & Rachlinski, J. J., 'They saw a protest: cognitive illiberalism and the speech-conduct distinction', *Stanford Law Review* 64, 851–906 (2012).

Tajfel, H., 'Experiments in intergroup discrimination', *Scientific American* 223(5), 96–102 (1970).

Steffens, N. K. & Haslam, S. A., 'Power through "Us": Leaders' use of we-referencing language predicts election victory', *PLoS One* 8(10): e77952 (2013). (https://doi.org/10.1371/journal.pone.0077952)

Alban, L. S. & Groman, W. D., 'Neurotic anxiety, pronoun usage and stress', *Journal of Clinical Psychology* 32(2), 393–9 (1976).

Pennebaker, J. W. & Chung, C. K., 'Computerized text analysis of Al-Qaeda transcripts', in Krippendorff, K. & Bock, M. (Eds.), *A content analysis reader* (Thousand Oaks, CA: Sage, 2008).

Bloom, M., 'The performance effects of pay dispersion on individuals and organizations'. *The Academy of Management Journal* 42(1), 25–40 (1999). (doi: 10.2307/256872)

Sean O'Brien, 'Forgiven: Sean Dyche offers Danny Drinkwater second chance at Burnley after Chelsea midfielder's nightclub brawl', *talkSport* website, 12 September 2019. (https://talksport.com/football/600589/sean-dyche-danny-drinkwater-burnley-chelsea-nightclub-brawl/)

Chapter 11: Undercover Influence

For the 'fox/hedgehog' quote and more of Archilochus' work, see: West, M. L. (Ed.), *Iambi et elegi Graeci: Ante Alexandrum cantati*, Vol. 1: Archilochus, Hipponax, Theognidea, 2nd ed., (Oxford: Oxford University Press, 1971, 1989).

Berlin, I., *The hedgehog and the fox: An essay on Tolstoy's view of history* (London: Weidenfeld & Nicolson, 1953).

Kahneman, Daniel, *Thinking, fast and slow* (New York: Farrar, Straus & Giroux, 2011).

For more on the perspective of autism and psychosis representing opposite ends of 'bounded self' spectrum, see: Crespi, B. & Badcock, C., 'Psychosis and autism as diametrical disorders of the social brain', *Behavioral and Brain Sciences* 31(3), 241–60 (2008) (doi:10.1017/S0140525X08004214); and Crespi, B. & Dinsdale, N., 'Autism and psychosis as diametrical disorders of embodiment', *Evolution, Medicine, and Public Health* (1), 121–38, 2019. (doi:10.1093/emph/eoz021)

Kat McGowan, 'The new quitter', *Psychology Today*, 1 July 2010. (https://www.psychologytoday.com/gb/articles/201007/the-new-quitter)

Ferguson, Alex & Michael Moritz, *Leading* (London: Hodder & Stoughton, 2015).

For more on the construct of cognitive complexity, see: Bieri, J., 'Cognitive complexity-simplicity and predictive behavior', *Journal of Abnormal and Social Psychology* 51, 263–8 (1955); Vannoy, J. S., 'Generality of cognitive complexity-simplicity as a personality construct', *Journal of Personality and Social Psychology* 2, 385–96 (1965); and Rafaeli-Mor, E., Gotlib, I. H. & Revelle, W., 'The meaning and measurement of self-complexity', *Personality and Individual Differences* 27, 341–56 (1999).

Hirsh, J. B., DeYoung, C. G., Xu, X. & Peterson, J. B., 'Compassionate liberals and polite conservatives: Associations of agreeableness with political ideology and moral values'. *Personality and Social Psychology Bulletin* 36(5): 655–64 (2010).

For more on differences in cognitive style between liberals and conservatives, see: Tetlock, P. E., 'Accountability and complexity of thought', *Journal of Personality and Social Psychology* 45(1), 74–83 (1983); Talhelm, T., Haidt, J., Oishi, S., Zhang, X., Miao, F. F. & Chen, S., 'Liberals think more analytically (more "WEIRD") than conservatives', *Personality and Social Psychology Bulletin* 41(2), 250–67 (2015); and Yilmaz, O. & Saribay, S. A., 'An attempt to clarify the link between cognitive style and political ideology: A non-western replication and extension', *Judgment and Decision Making* 11(3), 287–300 (2016).

For more on differences in personality style between liberals and conservatives, see: Carney, D. R., Jost, J. T., Gosling, S. D. & Potter, J., 'The secret lives of liberals and conservatives: Personality profiles, interaction styles, and the things they leave behind', *Political Psychology* 29(6), 807–40 (2008).

Jost, J. T., Napier, J. L., Thórisdóttir, H., Gosling, S. D., Palfai, T. P. & Ostafin, B., 'Are needs to manage uncertainty and threat associated with political conservatism or ideological extremity?', *Personality and Social Psychology Bulletin*, 33(7), 989–1007 (2007).

For more on both the cognitive and motivational factors underpinning religious and political extremism, see: Zmigrod, L., Rentfrow, P. J. & Robbins, T. W., 'Cognitive inflexibility predicts extremist attitudes', *Frontiers in Psychology* 10, 989 (2019) (doi: 10.3389/fpsyg.2019.00989); Zmigrod, L., Rentfrow, P. J. & Robbins, T. W., 'The partisan mind: Is extreme political partisanship related to cognitive inflexibility?' *Journal of Experimental Psychology: General* 149(3), 407–18 (2020) (https://doi.org/10.1037/xge0000661); Brandt, M. J. & Reyna, C., 'The role of prejudice and the need for closure in religious fundamentalism', *Personality and Social Psychology Bulletin* 36, 715–25 (2010) (doi: 10.1177/ 0146167210366306); Brewer, M. B., 'The psychology of prejudice: Ingroup love and outgroup hate?', *Journal of Social Issues* 55, 429–44 (1999) (doi: 10.1111/0022-4537.00126); and Tajfel, H. & Turner, J. C., 'An integrative theory of intergroup conflict', *The Social Psychology of Intergroup Relations* 33, 74 (1979).

Ingersoll, Ralph, *Report on England, November 1940* (New York: Simon and Schuster, 1940).

Mimi Launder & Greg Evans, '12 things politicians have said about Brexit which look a little naive now', *Indy 100*, 20 March 2019. (https://www.

indy100.com/article/no-deal-brexit-adequate-food-boris-johnson-david-davis-dominic-raab-8463121)

For more on task-oriented leadership, see: Fiedler, Fred E., 'A theory of leadership effectiveness' (New York: McGraw-Hill, 1967).

For more on leadership in general, see: Alexander Haslam, Stephen D. Reicher & Michael J. Platow, *The new psychology of leadership: Identity, influence and power* (Hove: Psychology Press, 2011).

Alison Rourke, 'Greta Thunberg responds to Asperger's critics: "It's a superpower"', *Guardian*, 2 September 2019. (https://www.theguardian.com/environment/2019/sep/02/greta-thunberg-responds-to-aspergers-critics-its-a-superpower)

Samantha Croal, 'Nicola Sturgeon slams Catherine Calderwood's actions: "There isn't one rule for her and another for everyone else"', *Daily Record*, 6 April 2020. (https://www.dailyrecord.co.uk/news/scottish-news/nicola-sturgeon-says-there-isnt-21819656)

Clare Foges, 'Dominic Cummings makes a mockery of his own coronavirus policy', *The Times*, 25 May 2020. (https://www.thetimes.co.uk/article/dominic-cummings-makes-a-mockery-of-his-own-coronavirus-policy-78w09n5h2)

Christopher Hope, 'The story behind "Stay Home, Protect the NHS, Save Lives" – the slogan that was "too successful"', *Telegraph*, 1 May 2020. (https://www.telegraph.co.uk/politics/2020/05/01/story-behind-stay-home-protect-nhs-save-lives/)

Dan Sabbagh, 'The week of "staying alert" that left the government in a daze', *Guardian*, 15 May 2020. (https://www.theguardian.com/world/2020/may/15/the-week-of-staying-alert-that-left-the-government-in-a-daze-coronavirus-boris-johnson)

Westaby, Stephen, *The knife's edge: The heart and mind of a cardiac surgeon* (London: Mudlark, 2019).

Nashef, Samer, *The angina monologues: Stories of surgery for broken hearts* (London: Scribe, 2019).

Whitehead, Alfred North, *Science and the modern world*, Lowell lectures, 1925, p. 54 (New York: The Free Press, 1967).

Kevin Rawlinson, ' "This enemy can be deadly": Boris Johnson invokes wartime language', *Guardian*, 17 March 2020. (https://www.theguardian.

com/world/2020/mar/17/enemy-deadly-boris-johnson-invokes-wartime-language-coronavirus)

For a full transcript of the Queen's address to the nation during the Coronavirus crisis, see: Telegraph Reporters, the Queen's coronavirus speech transcript: 'We will succeed and better days will come', *Telegraph*, 5 April 2020. (https://www.telegraph.co.uk/news/2020/04/05/queens-coronavirus-speech-full-will-succeed-better-days-will/).

Rachel Schraer, 'Coronavirus: Mental health of NHS staff at long-term risk', *BBC News* website, 15 May 2020. (https://www.bbc.co.uk/news/health-52528619)

Paul Thompson, 'Leicester City midfielder Hamza Choudhury apologises after racist and sexist tweets he wrote when he was 15 and 16 resurface', *Mail Online*, 29 April 2019. (https://www.dailymail.co.uk/news/article-6973537/Leicester-City-midfielder-Hamza-Choudhury-apologises-racist-sexist-tweets-resurface.html)

Dave Schilling, 'How Donald Trump tried to assimilate into Earth culture and failed', *Guardian*, 2 August 2016. (https://www.theguardian.com/us-news/2016/aug/02/donald-trump-eats-kfc-knife-fork)

Matt Flegenheimer, 'What democracy scholars thought of Trump's Bible photo op', *New York Times*, 2 June 2020. (https://www.nytimes.com/2020/06/02/us/politics/trump-holds-bible-photo.html)

Ottati, V., Rhoads, S. & Graesser, A. C., 'The effect of metaphor on processing style in a persuasion task: A motivational resonance model', *Journal of Personality and Social Psychology* 77(4), 688–97 (1999). (https://doi.org/10.1037/0022-3514.77.4.688)

Bernard Hickey, 'We must go hard and we must go early', *Newsroom*, 16 March 2020. (https://www.newsroom.co.nz/2020/03/14/1083045/we-must-go-hard-and-we-must-go-early)

Alistair Campbell, 'Jacinda Ardern's coronavirus plan is working because, unlike others, she's behaving like a true leader', *Independent*, 12 April 2020. (https://www.independent.co.uk/voices/coronavirus-new-zealand-jacinda-ardern-cases-deaths-leadership-a9460591.html)

UK and New Zealand death counts from COVID-19 taken from: https://www.worldometers.info/coronavirus/country/new-zealand/ (accessed 20 April 2020).

Saj Chowdhury, 'Christian Coleman defends himself against criticism over missed doping tests', *BBC Sport* website, 28 September 2019. (https://www.bbc.co.uk/sport/athletics/49867141)

Julia Carrie Wong, 'Elon Musk: Pedo guy insult was "not classy" but not meant literally', *Guardian*, 4 December 2019. (https://www.theguardian.com/technology/2019/dec/04/elon-musk-pedo-guy-insult-was-not-classy-but-not-meant-literally)

Read Sacha Baron Cohen's scathing attack on Facebook in full: 'Greatest propaganda machine in history', *Guardian*, 22 November 2019. (https://www.theguardian.com/technology/2019/nov/22/sacha-baron-cohen-facebook-propaganda)

For more on the Northern Ireland conflict, see: McKittrick, David & McVea, David, *Making sense of the Troubles: A history of the Northern Ireland conflict* (London: Penguin, 2012).

Alex Brownsell, 'KFC: A very fcking clever campaign', *Campaign*, 21 November 2018. (https://www.campaignlive.co.uk/article/kfc-fcking-clever-campaign/1498912)

"Imagine yourself alone in a park in Belfast, circa 1985 . . ." A similar story is told by the former MI5 operative Tom Marcus in his book *Soldier Spy* (London: Michael Joseph, 2016).

Chapter 12: Redrawing the Lines

Michael Hodges, 'Is Piers Morgan a psychopath?', *Radio Times*, 20 June 2019. (https://www.radiotimes.com/news/tv/2019-06-20/is-piers-morgan-a-psychopath/)

Dutton, Kevin & Andy McNab, *The good psychopath's guide to success: How to use your inner psychopath to get the most out of life* (London: Bantam, 2014).

Dutton, Kevin, *The wisdom of psychopaths: Lessons in life from saints, spies and serial killers* (London: Arrow, 2012).

Dutton, K., 'Would you vote for a psychopath? Research shows that some of the component traits of psychopathy can help leaders succeed', *Scientific American Mind*, September 2016, 50–5.

For more on the principle of accentuation, see: Krueger, J. & Clement, R. W., 'Memory-based judgments about multiple categories: A revision and extension of Tajfel's accentuation theory', *Journal of Personality and Social Psychology* 67(1), 35–47 (1994); Tajfel, H., 'Quantitative judgment in social perception', *British Journal of Psychology* 50, 16–29 (1959); Tajfel, H., 'Cognitive aspects of prejudice', *Journal of Social Issues* 25, 79–97 (1969); and Tajfel, H. & Wilkes, A. L., 'Classification and quantitative judgment', *British Journal of Psychology* 54, 101–14 (1963).

Oliver Milne, 'Coronavirus testing should include "virtually everybody" to stop spread, says Tony Blair' *Daily Mirror Online*, 29 March 2020. (https://www.mirror.co.uk/news/politics/coronavirus-testing-should-include-virtually-21774137)

Aristotle, 'Categories', in Barnes, J. (Ed.), *The complete works of Aristotle*, translated by Ackrill, J. L. (Princeton: Princeton University Press, 2014).

For George Lakoff quote, see: Janicki, Karol, 'The ever-stifling essentialism: Language and conflict in Poland (1991–1993)', in Hubert Cuyckens, Thomas Berg, René Dirven & Klaud-Uwe Panther (Eds), *Motivation in language: Studies in honor of Günter Radden*, pp. 273–95 (Amsterdam: John Benjamins, 2003).

For more on essentialism, see: Lakoff, George & Mark Johnson, *Philosophy in the flesh* (New York: Basic Books, 1999); Cartwright, R. L., 'Some remarks on essentialism', *The Journal of Philosophy* 65(20), 615–26 (1968) (doi:10.2307/2024315); and Bastian, B. & Haslam, N., 'Psychological essentialism and stereotype endorsement', *Journal of Experimental Social Psychology* 42(2), 228–35 (2006). (doi:10.1016/j.jesp.2005.03.003)

For more on the psychological factors involved in attitude change, see: Persuasion. Psychology Research and Reference Net (http://psychology.iresearchnet.com/social-psychology/social-influence/persuasion/); and Petty, R. E., 'Attitude change', in Tesser, A. (Ed.), *Advanced social psychology*, pp. 195–255 (New York: McGraw-Hill, 1995).

Graeme Bruce, 'Corona beer brand is impacted by the coronavirus news', *YouGov* website, February 2020. (https://today.yougov.com/topics/food/articles-reports/2020/02/26/first-it-was-joke-corona-beer-really-suffering-cor)

For the lemon/lime 7-Up study, see: Michelle Meyer, 'Good things come in new packages', *Arrive* (November/December, 2007). (https://www.michelemode.com/wp-content/uploads/2013/09/packages.arrive.pdf)

For more on the science of food and how our perception of it is cross-modal and shaped not just by taste but by all our senses, see: Spence, Charles, *Gastrophysics: The new science of eating* (London: Penguin, 2017).

For more on in-group bias and how the right kind of contact with other groups can help reduce it, this classic is still the best book out there: Sherif, M., Harvey, O. J., White, B. Jack, Hood, William R. & Sherif, Carolyn W., *Intergroup conflict and cooperation: The robbers cave experiment* (Norman, OK: The University Book Exchange, 1954, 1961).

Abrams, D., Wetherell, M., Cochrane, S., Hogg, M. & Turner, J. C., 'Knowing what to think by knowing who you are: self-categorization and the nature of norm formation, conformity and group polarization', *British Journal of Social Psychology* 29(2), 97–119 (1990). (https://doi.org/10.1111/j.2044-8309.1990.tb00892.x)

For the full story on Dorothy Martin and the Seekers, see: Festinger, Leon, Riecken, Henry & Schachter, Stanley, *When prophecy fails: A social and psychological study of a modern group that predicted the end of the world* (Minneapolis: University of Minnesota Press, 1956).

For an overview of cognitive dissonance theory and related accounts of cognitive consistency, see: Festinger, Leon, *A theory of cognitive dissonance* (Stanford, CA: Stanford University Press, 1957); Festinger, Leon & Carlsmith, J. M., 'Cognitive consequences of forced compliance', *Journal of Abnormal and Social Psychology* 58, 203–10 (1959); and Cooper, J. & Fazio, R. H., 'A new look at dissonance theory', in L. Berkowitz (Ed.), *Advances in experimental social psychology*, Vol. 17, 229–64 (Orlando, FL: Academic Press, 1984).

Temple, Olivia & Temple, Robert (translators), *Aesop: The complete fables* (New York: Penguin Classics, 1998).

Rebecca Perring, 'REVEALED: The EU's top ten pointless decisions the UK can now get rid of', *Express*, 25 June 2016. (https://www.express.co.uk/news/world/586742/European-Union-barmy-decisions-rules-regulations-Britain-EU)

Richard Allen Greene, 'Brexit poll: Six months on, Brits stand by EU referendum decision', *CNN World*, 19 December 2016. (https://edition.cnn.com/2016/12/19/europe/cnn-brexit-poll/index.html)

Lionel Barber, 'Fake news in the post-factual age', lecture to Oxford alumni festival, Oxford University, *Financial Times*, 16 September 2017. (https://www.ft.com/content/c8c749e0-996d-11e7-b83c-9588e51488a0)

Jonathan Freedland, 'The great divide of our times is not left v right, but true v false', *Guardian*, 20 April 2018. (https://www.theguardian.com/commentisfree/2018/apr/20/trump-us-syria-truth-tribal-robert-mueller-white-helmets-factse)

For Scottie Nell Hughes quote, see: Sofia Lotto Persio, 'Trump vows to "strengthen voting procedures" after ordering probe into already debunked voter fraud', *International Business Times*, 25 January 2017. (https://www.ibtimes.co.uk/trump-strengthen-voting-procedures-following-probe-into-already-debunked-voter-fraud-1603046)

For the Dmitry Kiselyov quote and a discussion of the propaganda arms race between Russia and the West, see: Mark Galeotti, 'The west is too paranoid about Russia's information war', *Guardian*, 7 July 2015. (https://www.theguardian.com/world/2015/jul/07/russia-propaganda-europe-america)

For more on the emergence of post-truth values, see: John Keane, 'Post-truth politics and why the antidote isn't simply "fact-checking" and truth', *The Conversation*, 23 March 2018. (https://theconversation.com/post-truth-politics-and-why-the-antidote-isnt-simply-fact-checking-and-truth-87364); for more on societal precursors to the post-truth climate, see: John Keane, *Democracy and media decadence* (Cambridge: Cambridge University Press, 2013).

Bourdieu, Pierre, *Distinction: A social critique of the judgment of taste*, translated by Richard Nice, p. 479 (Cambridge MA: Harvard University Press, 1984).

Sally Weale & Frances Perraudin, 'Jazz hands at Manchester University: The calm behind the storm', *Guardian*, 5 October 2018. (https://www.theguardian.com/society/2018/oct/05/jazz-hands-at-manchester-university-the-calm-behind-the-storm)

Shari Miller, 'Theatres will BAN the phrase "ladies and gentlemen" under new guidelines from actors' union Equity', *Mail Online*, 3 November 2019. (https://www.dailymail.co.uk/news/article-7644549/Ladies-gentlemen-banned-theatre-actor-union-Equitys-new-gender-neutral-guide.html)

Helen Coffey, 'EasyJet swaps "ladies and gentlemen" for more gender inclusive language', *Independent*, 24 December 2019. (https://www.independent.

co.uk/travel/news-and-advice/easyjet-gender-neutral-inclusive-language-trans-ladies-gentlemen-a9258961.html)

For a little bit more on quarter tones, see: A. Eric Heukeshoven, 'West meets East – Notation & playback of quarter tone music using Sibelius', *Of Note*, 11 June 2012. (https://www.rpmseattle.com/of_note/west-meets-east-notation-playback-of-quarter-tone-music-using-sibelius/)

Mere Abrams, '46 terms that describe sexual attraction, behaviour, and orientation', *Healthline Media*, 10 December 2019. (https://www.healthline.com/health/different-types-of-sexuality)

Simon Hattenstone, 'Anthony Ekundayo Lennon on being accused of "passing" as a black man: "It felt like an assassination"', *Guardian*, 7 September 2019. (https://www.theguardian.com/world/2019/sep/07/anthony-lennon-theatre-director-accused-of-passing-as-black-interview-simon-hattenstone)

'Dutch man, 69, who "identifies as 20 years younger" launches legal battle to change age', *Telegraph*, 7 November 2018. (https://www.telegraph.co.uk/news/2018/11/07/dutch-man-69-identifies-20-years-younger-launches-legal-battle/)

Sharon Otterman, 'Sandy Hook conspiracy theorist loses to father of 6-year-old victim over hoax', *New York Times*, 18 June 2019. (https://www.nytimes.com/2019/06/18/nyregion/sandy-hook-victim-court-ruling.html).

Christopher Nehring, 'Coronavirus and the plague: The disease of viral conspiracy theories', *DW Akadamie*, 10 March 2020. (https://www.dw.com/en/coronavirus-and-the-plague-the-disease-of-viral-conspiracy-theories/a-52700321)

Lora Jones & Jane Wakefield, 'Coronavirus: Supermarkets limit shoppers as rules tighten', *BBC News* website, 24 March 2020. (https://www.bbc.co.uk/news/business-52022240)

Everett, J. A. C., Colombatto, C., Chituc, V., Brady, W. J. & Crockett, M., 'The effectiveness of moral messages on public health behavioral intentions during the COVID-19 pandemic', *PsyArXiv Preprints*, 20 March 2020. (https://doi.org/10.31234/osf.io/9yqs8)

Leo Kelion, 'Coronavirus: Millions yet to receive UK government's Covid-19 text', *BBC News* website, 25 March 2020. (https://www.bbc.co.uk/news/technology-52037573)

Buttrick, N., Moulder, R. & Oishi, S., 'Historical change in the moral foundations of political persuasion', *Personality and Social Psychology Bulletin* (2020). (https://doi.org/10.1177/0146167220907467)

Martin Robinson, 'West Midlands chief constable blasts claims Britain is becoming a "police state" … but he and Grant Shapps are forced to admit some officers have gone "further than they should" during virus lockdown', *Daily Mail*, 31 March 2020. (https://www.dailymail.co.uk/news/article-8170863/Police-warned-against-heavy-handed-lockdown-enforcement-five-days-ago.html)

Guidance issued on new police powers, *Police Federation* website, 26 March 2020. (https://www.polfed.org/news-media/latest-news/2020/guidance-issued-on-new-police-powers/)

Laura Delizonna, 'High-performing teams need psychological safety. Here's how to create it', *Harvard Business Review*, 24 August 2017. (https://hbr.org/2017/08/high-performing-teams-need-psychological-safety-heres-how-to-create-it)

Roger Dean Duncan, 'Avoid the blame game: Be accountable for accountability', *Forbes*, 4 May 2018. (https://www.forbes.com/sites/rodgerdeanduncan/2018/05/04/avoid-the-blame-game-be-accountable-for-accountability/#96757912b229)

For more on the productive power of harnessing superordinate group identity, see the aforementioned Muzafer Sherif et al., *Intergroup conflict and cooperation: The robbers cave experiment* (Norman, OK: The University Book Exchange, 1954, 1961).

For the seating-plan study, see: Johnson, David W. & Johnson, Frank P., *Joining together: Group theory and group skills* (Englewood Cliffs, NJ: Prentice Hall, 1975).

For the University of Delaware football stadium study, see: Gaertner, S. L., Dovidio, J. F. & Bachman, B. A., 'Revisiting the contact hypothesis: The induction of a common in-group identity', *International Journal of Intercultural Relations* 20, 271–90 (1996).

Postscript: The Wisdom of Radicals

Jeff D. Gorman, 'Emirati man sues Marriott over Islamic State accusation', *Courthouse News*, 25 May 2017. (https://www.courthousenews.com/emirati-man-sues-marriott-isis-pledge-accusation/)

Steph Solis, 'Pokémon players shot at in Palm Coast, Fla', *USA Today*, 18 July 2016. (https://eu.usatoday.com/story/tech/gaming/2016/07/16/pokmon-trainers-shot-palm-coast-florida/87207116/)

John Scheerhout, 'Blind man Tasered by police after his walking cane was mistaken for gun', *Manchester Evening News*, 24 February 2017. (https://www.manchestereveningnews.co.uk/news/greater-manchester-news/blind-man-tasered-police-after-12652337)

Kashmira Gander, 'Michael Brown shooting: What happened in Ferguson?' *Independent*, 11 August 2015. (https://www.independent.co.uk/news/world/americas/michael-brown-shooting-what-happened-in-ferguson-10450257.html)

McSpadden, Lezley & Lyah Beth LeFlore, *Tell the truth and shame the devil: The life, legacy, and love of my son Michael Brown* (New York: Regan Arts, 2016).

Payne, B. K., 'Prejudice and perception: The role of automatic and controlled processes in misperceiving a weapon', *Journal of Personality and Social Psychology* 81, 181–92 (2001).

Correll, J., Urland, G. L. & Ito, T. A., 'Event-related potentials and the decision to shoot: The role of threat perception and cognitive control', *Journal of Experimental Social Psychology* 42, 120–8 (2006).

Sagar, H. A. & Schofield, J. W., 'Racial and behavioral cues in black and white children's perceptions of ambiguously aggressive acts', *Journal of Personality and Social Psychology* 39(4), 590–8 (1980).

For more on the functional utility of stereotypes as a quick and dirty means of survival, see: Macrae, C. N., Milne, A. B. & Bodenhausen, G. V., 'Stereotypes as energy-saving devices: A peek inside the cognitive toolbox', *Journal of Personality and Social Psychology* 66(1), 37–47 (1994). (doi: 10.1037/0022-3514.66.1.37)

For a wider and more in-depth exploration of the science of stereotypes in general, see: Macrae, C. Neil, Charles Stangor & Miles Hewstone (Eds), *Stereotypes and stereotyping* (New York: Guilford, 1996).

For more on neuronal activity and the all-or-none principle, see: Kendra Cherry, 'All-or-none law for nerves and muscles', *Very Well Mind*, 10 April 2020. (https://www.verywellmind.com/what-is-the-all-or-none-law-2794808)

James, William, *Some problems of philosophy*, p. 78 (Lincoln, NE: University of Nebraska Press, 1911, 1996).

For the source of the Albert Einstein quote 'Everything should be made as simple as possible but no simpler', see: 'On the method of theoretical physics', Herbert Spencer Lecture, delivered at Oxford (10 June 1933); also published in *Philosophy of Science* 1(2), April 1934, 163–169.

Appendix III: A brief history of frames

Kant, Immanuel, *Critique of pure reason*, 1781, translated by Guyer, P. & Wood, A. W. (Cambridge: Cambridge University Press, 1997).

Nevid, J. S., 'Kant, cognitive psychotherapy, and the hardening of the categories', *Psychology and Psychotherapy: Theory, Research and Practice* 80(4): 605–615 (2007). (doi:10.1348/147608307X204189)

Piaget, Jean, *The language and thought of the child* (London: Routledge, 1923, 2002).

Bartlett, F. C., *Remembering: A study in experimental and social psychology* (Cambridge: Cambridge University Press, 1932, 1997).

Minsky, M., 'A framework for representing knowledge', in *The psychology of computer vision*, Winston, P. H. (Ed.) (New York: McGraw-Hill, 1975).

Rumelhart, D. E., 'Schemata: The building blocks of cognition', in Spiro, R. J., Bruce, B. C. & Brewer, W. F. (Eds), *Theoretical issues in reading comprehension* (Hillsdale, NJ: Erlbaum, 1980).

Schank, Roger & Robert P. Abelson, *Scripts, plans, goals and understanding: An inquiry into human knowledge structures* (New Jersey: Erlbaum, 1977).

For more on the psychology of humour, see: Jimmy Carr & Lucy Greeves, *The naked jape: Uncovering the hidden world of jokes* (London: Penguin, 2006).

Lakoff, George & Mark Johnson, *Metaphors we live by* (Chicago: University of Chicago Press, 1980).

Matt Bai, 'The framing wars', *New York Times*, 17 July 2005. (https://www.nytimes.com/2005/07/17/magazine/the-framing-wars.html)

For the *Fawlty Towers* restaurant clip ('The Germans', Series 1, Episode 6, original air date 24 October 1975), see: https://www.youtube.com/watch?v=yfl6Lu3xQW0.

Appendix V: The three evolutionary stages of black-and-white thinking

McMenamin, M. A. S., 'Palaeocological feedback and the Vendian-Cambrian transition', *Trends in Ecology and Evolution* 3(8), 205–8 (1988). (doi:10.1016/0169-5347(88)90008-0)

Dunbar, Robin, *Grooming, gossip, and the evolution of language* (Cambridge, MA: Harvard University Press, 1996).

Lehmann, J. & Boesch, C., 'To fission or to fusion: Effects of community size on wild chimpanzee (*Pan troglodytes verus*) social organisation', *Behavioral Ecology and Sociobiology* 56, 207–16 (2004). (doi.org/10.1007/s00265-004-0781-x)

Symington, M. M., 'Fission-fusion social organization in Ateles and Pan', *International Journal of Primatology* 11, 47–61 (1990). (doi.org/10.1007/BF02193695)

Whiten, A. & Erdal, D., 'The human socio-cognitive niche and its evolutionary origins', *Philosophical Transactions of the Royal Society of London. Series B, Biological Sciences* 367(1599), 2119–29 (2012). (doi.org/10.1098/rstb.2012.0114)

Bird, D. W., Bird, R. B., Codding, B. F. & Zeanah, D., 'Variability in the organization and size of hunter-gatherer groups: Foragers do not live in small-scale societies', *Journal of Human Evolution* 131, 96–108 (2019).

For more on the brain's evolved ability to detect shifts in coalitional alliances, see: Kurzban, R., Tooby, J. & Cosmides, L., 'Can race be erased? Coalitional computation and social categorization', *Proceedings of the National Academy of Sciences* 98, 15387–92 (2001).

Stiner, M. C. & Kuhn, S. L., 'Changes in the "connectedness" and resilience of Paleolithic societies in Mediterranean ecosystems', *Human Ecology* 34, 693–712 (2006). (doi.org/10.1007/s10745-006-9041-1)

Schick K. D., 'A comparative perspective on Paleolithic cultural patterns', in Akazawa, T., Aoki, K. & Bar-Yosef, O. (Eds), *Neandertals and Modern Humans in Western Asia* (Boston, MA: Springer, 2002). (doi.org/10.1007/0-306-47153-1_29)

Burkart, J. M., Brügger, R. K. & van Schaik, C. P., 'Evolutionary origins of morality: Insights from non-human primates', *Frontiers in Sociology* 3, 17 (2018). (doi:10.3389/fsoc.2018.00017)

Appendix VI: Black-and-white thinking through the centuries

Baumeister, R. F., 'How the self became a problem: A psychological review of historical research', *Journal of Personality and Social Psychology* 52(1), 163–176 (1987).

Singman, Jeffrey L., *Daily life in medieval Europe* (Westport, CT: Greenwood, 1999).

Nisbet, Robert, *The social philosophers: Community and conflict in Western thought* (New York: Crowell, 1973).

Zafirovski, Milan, *The enlightenment and its effects on modern society* (New York: Springer, 2010).

Lewis, Pericles, *The Cambridge introduction to modernism* (Cambridge: Cambridge University Press, 2007).

Cahoone, Lawrence E. (Ed.), *From modernism to postmodernism: An anthology* (Malden, MA: Blackwell, 2003).

Daunton, Martin (Ed.), *The organisation of knowledge in Victorian Britain* (Oxford: Oxford University Press, 2005).

Lyotard, Jean-François, *The postmodern condition: A report on knowledge* (Manchester: Manchester University Press, 1984).

Musil, Robert & Burton Pike (Eds), *The man without qualities*, translated by Sophie Wilkins & Burton Pike (New York: Knopf, 1995).

Appendix VII: The essentials of essentialism

Paul Waugh, 'Boris Johnson's blame game on Covid-19 finally begins to reveal itself', *Huffington Post*, 28 May 2020. (https://bit.ly/2ApJN2R)

Acknowledgements

There is nothing black and white about writing a book like this. It may seem that way from a distance. There is only one name on the cover. But the reality is rather different. Writing a book like this is like working with stained glass. It is light that gives it life. Such light is extraneous to the artistic process itself. But it is the electromagnetic heartbeat of the oeuvre.

Over the course of the last three years there have been a great many people who've formed the narrative heartbeat of this book: who've illuminated my footsteps, enlightened my thinking, shone a critical torch into the impenetrable fog of ideological disillusionment and brightened up the long, dark days when words, inspiration and pay cheques failed to materialize.

Editors, someone once said, are the traffic cops of the arts. If that is true then I've been booked many times. Stop lights, hazard warning lights and blue flashing lights have all been part of the wider editorial backdrop to *Black-and-White Thinking* and I am eternally grateful that Andrea Henry at Bantam Press and Colin Dickerman at Farrar, Straus and Giroux wrote me out the tickets when they did. Artistic pile-ups awaited us at every turn. But somehow we managed to avoid them. Andrea and Colin tailed me all the way.

Other lights shone, too, in the creative firmament. Mentioning them all would put me way and above an already exceeded word count . . . and earn me another ticket. There are, quite simply, too many. Here, though, are a few that really dazzled: Dominic Abrams, Mike Anderson, Tony 'Frog' Barlow, Joey Barton, Mark Bassett, Alfie Best, Tony Blair, Carole Bloch, Alastair Campbell, Ginny Carter, Robert Cialdini, Sarah Classick, Lord Sebastian Coe, Ian Collins, John Collins, Sir Ivor Crewe, Lawrence Dallaglio, Jules

Davidoff, Sir Ray Davies, Richard Dawkins, Susie Dent, Elizabeth Dobson, Sean Dyche, George Ellis, Angela Entwistle, Sir Ranulph Fiennes, Tory Finch, Katharina Festner, Alex Fisher, Ashley Giles, Gavin Hewitt, Peter and Ana Holland, Diane Houston, Eddie Howe, Eddy Jennings, Hugh Jones, (the late) Terry Jones, Neil Kerly, Gilda Kiai, Lynn Kimsey, Arie Kruglanski, Gemma Levine, John Lloyd, Colin MacLeod, Kellie Maloney, Ben Matthews, Andy McNab, Amanda Moon, Ben Morgan, Piers Morgan, Jeff Mostyn, Rob Murray, Samer Nashef, Lisa Oakes, The OCEAN Lab (Dept of Experimental Psychology, University of Oxford), Ronnie O'Sullivan, Nick Pullen, Sally Pullen, Sir Philip Pullman, Debi Roberson, Martin Roberts, Eleanor Rosch, Francis Rossi, Bill Scott-Kerr, Raj Sehgal, Ben Shephard, Laura Sherlock, Paul Sinton-Hewitt, Claire Smith, Anna Soderstrom, Annabel Songco, Micky Speight, Iwan Thomas, Richard Thompson, Harvey and Alice Thorneycroft, Henry Thorpe, Eleanor Updegraff, John Virgo, Gregg Wallace, Stephen Westaby, Matt White, Kit and Eleyne Williams, Andrea Woerle, and Sally Wray.

Time for a terrible transition. A publisher and an agent stop for a drink after work. 'I don't understand,' complains the agent. 'People take an instant dislike to me when they find out I'm an agent. Why would they do that?'

'Maybe,' the publisher suggests, 'it just saves time.'

Thank you, Patrick Walsh, for not being that agent. And for your continued support and friendship over the years.

Three other people are deserving of special mention.

First, Wing Commander Keith Dear, whose enthusiasm for sourcing the material for this book remained constant throughout and unstinting right to the end. No commentary was deemed too irrelevant. No reference too obscure. He even went to the trouble of putting his own name forward for the acknowledgements. Beyond the call of duty.

Second, my copyeditor, Mari Roberts, whose effortless professionalism and understated intolerance of the pompous, verbose and grandiose shone through once again in the finished manuscript at a particularly testing time.

And finally, my immeasurably superior other half, Elaine: 'The brains of the outfit', as a certain Eddy Jennings once observed. Elaine has been in the Kevin Dutton business for over twenty years.

Miraculously, we're still trading.

Index

Illustration Credits

CPSIA information can be obtained
at www.ICGtesting.com
Printed in the USA
LVHW041209190122
708804LV00002B/3

9 781250 829450